Optical Waveguide Sensing and Imaging

edited by

Wojtek J. Bock

Université du Québec en
Outaouais, Gatineau, QC,
Canada

Israel Gannot

Tel-Aviv University, Tel-Aviv, Israel, and
George Washington University, Washington, DC, U.S.A.

and

Stoyan Tanev

Carleton University, Ottawa,
ON, Canada

 Springer

Published in cooperation with NATO Public Diplomacy Division

Proceedings of the NATO Advanced Study Institute on
Optical Waveguide Sensing and Imaging in Medicine, Environment,
Security and Defence
Gatineau, Québec, Canada
12–21 October 2006

A C.I.P. Catalogue record for this book is available from the Library of Congress.

ISBN 978-1-4020-6951-2 (PB)
ISBN 978-1-4020-6950-5 (HB)
ISBN 978-1-4020-6952-9 (e-book)

Published by Springer,
P.O. Box 17, 3300 AA Dordrecht, The Netherlands.

www.springer.com

Printed on acid-free paper

Optical W...aging

NATO Science for Peace and Security Series

This Series presents the results of scientific meetings supported under the NATO Programme: Science for Peace and Security (SPS).

The NATO SPS Programme supports meetings in the following Key Priority areas: (1) Defence Against Terrorism; (2) Countering other Threats to Security and (3) NATO, Partner and Mediterranean Dialogue Country Priorities. The types of meeting supported are generally "Advanced Study Institutes" and "Advanced Research Workshops". The NATO SPS Series collects together the results of these meetings. The meetings are co-organized by scientists from NATO countries and scientists from NATO's "Partner" or "Mediterranean Dialogue" countries. The observations and recommendations made at the meetings, as well as the contents of the volumes in the Series, reflect those of participants and contributors only; they should not necessarily be regarded as reflecting NATO views or policy.

Advanced Study Institutes (ASI) are high-level tutorial courses intended to convey the latest developments in a subject to an advanced-level audience

Advanced Research Workshops (ARW) are expert meetings where an intense but informal exchange of views at the frontiers of a subject aims at identifying directions for future action

Following a transformation of the programme in 2006 the Series has been re-named and re-organised. Recent volumes on topics not related to security, which result from meetings supported under the programme earlier, may be found in the NATO Science Series.

The Series is published by IOS Press, Amsterdam, and Springer, Dordrecht, in conjunction with the NATO Public Diplomacy Division.

Sub-Series

A.	Chemistry and Biology	Springer
B.	Physics and Biophysics	Springer
C.	Environmental Security	Springer
D.	Information and Communication Security	IOS Press
E.	Human and Societal Dynamics	IOS Press

http://www.nato.int/science
http://www.springer.com
http://www.iospress.nl

Series B: Physics and Biophysics

TABLE OF CONTENTS

PREFACE

Wojtek J. BOCK
Centre de recherche en photonique, Département d'informatique et d'ingénierie
Université du Québec en Outaouais, Québec, Canada

Israel GANNOT
Department of Biomedical Engineering, Faculty of Engineering
Tel-Aviv University, Tel-Aviv 69978, Israel
and
Department of Electrical and Computer Engineering,
School of Engineering and Applied Sciences
George Washington University, Washington, DC 20052, USA

Stoyan TANEV
Department of Systems and Computer Engineering
Faculty of Engineering and Design
Carleton University, Ottawa, Ontario, Canada K1S 5B6

The past decade has witnessed some major technological breakthroughs produced by the fusion of different disciplines. This trend is likely to develop in the future due to the recent significant advancements of fiber optics communications, photonics, biomedical and nano-technologies worldwide. In parallel with the communications and information technology revolution, fiber and waveguide optics sensor and imaging technologies have enjoyed an unseen technological maturity and revealed enormous potentials for a broad variety of new applications.

The content of this book is based on peer reviewed invited articles corresponding to some of the tutorial presentations delivered in the frame of the NATO Advanced Study Institute (ASI) "Optical Waveguide Sensing and Imaging in Medicine, Environment, Security and Defence" held in Gatineau, Québec, Canada, October 12–21, 2006. The objective of the ASI was to build a creative advanced research and learning environment by bringing together world experts, researchers, Ph.D. students and postdoctoral fellows from industry, university and government research organizations. The ASI explored a variety of aspects of research, development, and commercialization of existing and emerging optical waveguide, fiber, micro and nanophotonics imaging and sensing technologies, as well as their current and potential applications in the biomedical sciences, environmental monitoring, security and defence. The topics included in this volume can be subdivided in three major areas: i) advances in optical fiber sensing and imaging including the application of novel sensing mechanisms, ii) micro- and nanophotonics structures for imaging and sensing, and iii) biomedical optics modeling tools for sensing and imaging applications.

The editors are grateful to all supporting organizations and people that made this ASI possible. Our special thanks are directed towards the NATO Security through Science Program (Brussels, Belgium); Canadian Institutes of Health Research; Canadian Institute for Photonic Innovations; Université du Québec en

Outaouais, Gatineau, Québec; and Carleton University, Ottawa, Ontario, Canada. We thank the team of Vitesse Re-Skilling™ Canada and its President Arvind Chhatbar for partnering in setting the initial vision of the ASI, the program and financial management, as well as for providing the organizational infrastructure during the meeting. We are also grateful to the Canadian Department of Foreign Affairs and its Global Partnership Program, the International Science and Technology Center in Moscow (Russia) and the Science and Technology Center in Kiev (Ukraine), for supporting the participation of the Russian and the Ukrainian participants. We also thank the executive management team and, especially, Caroline Hamel, Product Line Manager, at FISO Technologies Inc. for the financial support of the ASI.

SENSOR APPLICATIONS OF FIBER BRAGG AND LONG PERIOD GRATINGS

TINKO EFTIMOV*
*Faculty of Physics, Plovdiv University "Paisii Hilendarski",
Plovdiv 4000, Bulgaria*

Abstract. The basic idea and theoretical description of fiber Bragg (FBG) and long period (LPG) gratings are presented. Sensitivity characteristics and methods of fabrication are considered. The various types of fiber optic grating sensors, multiplexing and interrogation techniques and domains of applications are presented and compared.

Keywords: Fibre optic sensors, fiber Bragg gartings (FBG), long period gratings (LPG), fiber optic sensor networks, structural health monitoring

1. Introduction

Since their first discovery (Hill, 1978) fiber gratings have greatly evolved and matured. Fiber gratings are structures consisting of a periodic per-turbation of the optical and/or geometrical properties of an optical fibre. Depending on the pitch Λ of the perturbation, fiber gratings fall into two distinct categories: short period gratings, known as fiber Bragg gratings (FBGs) and, long period gratings (LPGs) proposed about eight years later (Vengsarkar 1996). From the outset the unique spectral properties of fiber gratings determined their applications in the communication and in the sensor industry.

An enormous amount of publications exist on the fabrication technology and applications of the various types of fiber gratings and with the invention of microstructured fibers new unexpected opportunities and developments are constantly emerging. The present overview only deals with the sensor

---

* Tinko Eftimov, Department of Experimental Physics, Faculty of Physics, Plovdiv University "Paisii Hilendarski", Plovdiv 4000, Bulgaria, tinkoeftimov@hotmail.com.

W.J. Bock et al. (eds.), Optical Waveguide Sensing and Imaging, 1–23.
© 2008 *Springer.*

applications of fiber gratings. First the basic idea and properties of FBGs and LPGs are considered. Second, the fabrication techniques are outlined. Third, sensitivities of BFs and LPGs to different measurands are commented and examples of their sensor applications are presented. The last section, devoted to a comparison between the two types of gratings, shows, that they occupy different application niches and thus enormously widen the application fields of fibre optic sensor technogy.

2.   Fiber Bragg Gratings and Long Period Gratings

2.1. PHYSICAL PRINCIPLES

2.1.1. *FBGs*

We suppose a distributed periodic structure consisting of a series of transitions from a lower index n_0 to a higher index n as shown in Figure 1. This is a grating structure with a period Λ and an index modulation Δn. Waves reflected at each interface will interfere and for a given period Λ constructive interference due to phase matching will be observed only for a particular resonance wavelength such that $\lambda/2 = n_0\Lambda$. Such a structure is known as a Bragg grating and is characterized by its resonance Bragg wavelength $\lambda_B = 2n_0\Lambda$ at which the reflection (R) reaches a maximum and the transmission (T) – a minimum.

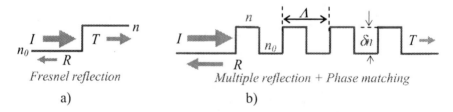

Figure 1. a) Fresnel reflection at the interface of two media. b) Reflection at a particular resonance wavelength λ_B from a periodic structure.

We now consider such a periodic structure induced along an optical fiber which uses total internal reflection to guide light waves. Its core has a lower refractive index n_1 compared to the cladding n_2 and we have wave structures, called "modes" travelling along the fiber. When the core supports only one mode the fiber is single-mode. This fundmental mode (FM) of the core has a propagation constant β_c such that $kn_2 < \beta_c = kn_c < kn_1$, where k is the free space wave number and n_c is the mode effective refractive index. The periodic structure causes the forward propagating core

FM to couple to a backward propagating core FM as well as to backward propagating higher-order cladding modes (HOCMs) shown in Figure 2, which have propagation constants β_{cl} such that $kn_{air} < \beta_{cl} = kn_{cl} < kn_2$, and n_{cl} is the cladding mode effective index.

Figure 2. The forward fundamental mode (FM) couples to a backward FM and to a multitude of higher-order cladding modes (HOCMs).

In a FBG the dominant effect is coupling to a backward core FM at λ_B. The basic idea beyond a FBG is thus counter-directional coupling. Depending on the optical Δn and geometric parameters Λ, various types of FBGs exist and the most important are summarized in Table 1.

TABLE 1. Main of types of FBGs

❖ **Weak**: small Δn	o **Strong**: large Δn
Reflectivity is 10 to 40% maximum	*Reflectivity is > 40%*
❖ **Uniform**: $\Delta n = const$	o **Apodized**: Δn varies
Strong side lobes	*Side lobes suppressed*
❖ **Regular**: $\Lambda = const$	o **Chirped**: Λ varies
Narrow bandwidth and low dispersion	*Broad bandwidth and high dispersion*
❖ **Straight FBGs**	o **Tilted**
Back-couples to a core mode	*Back-couples to radiation modes*

2.1.2. *LPGs*

In the case of LPGs the pitch length Λ is considerably greater and the periodic perturbation causes fractions of the forward propagating fundamental LP_{01} (HE_{11}) mode with a propagation constant $\beta_0 = kn_0$ to couple to forward propagating higher-order cladding LP_{0m} (HE_{1m}) modes with propagation constants $\beta_m = kn_m$, where n_0 and n_m are effective mode indices.

The higher-order modes suffer from higher attenuation and thus cause a minimum in the transmission of the fundamental mode. Unlike FBGs, no reflection is observed with LPGs. We thus have co-directional intermodal coupling between a core and a cladding mode caused by a periodic perturbation along the fiber and a phase matching condition.

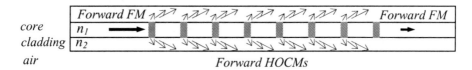

Figure 3. The forward propagating fundamental mode (FM) couples to forward propagating higher order cladding modes (HOCMs).

2.2. GRATING CHARACTERISTICS

The theoretical analysis of fiber gratings is performed on the basis of coupled mode differential equations (Erdogan, 1997) which describe evolution of the interacting mode amplitudes along the grating structure. These equations have different solutions for FBGs and LPGs and can be represented in a 2 × 2 matrix form from which intensities of transmitted and reflected light are calculated. The important parameters in these equations are the coupling coefficient κ and the detuning parameter δ which tells how

TABLE 2. Basic FBG and LPG characteristics

Parameter	FBG	LPG
δ	$\beta_c - \pi/\Lambda = 2\pi n_c (1/\lambda - 1/\lambda_D)$	$(\beta_c - \beta_m - 2\pi/\Lambda)/2$ $= \pi \Delta n_m (1/\lambda - 1/\lambda_D)$
$\lambda_{max} \approx \lambda_D$	$\lambda_D = \lambda_B = 2n_c\Lambda$	$\lambda_D = \lambda_m = \Delta n_m \Lambda$
Λ	$\approx 0,5\ \mu m$	$250 \div 750\ \mu m.$
$\dfrac{\Delta\lambda_0}{\lambda_{max}}$	$\dfrac{\Delta n}{n_c}\sqrt{1 + \left(\dfrac{\lambda_B}{\Delta n L}\right)^2}$	$2\alpha_{BRF}\dfrac{\Lambda}{L}\sqrt{1 - \left(\dfrac{\kappa L}{\pi}\right)^2}$
$\Delta\lambda_0$	< 1 nm (< 6 nm for chirped FBGs)	20–50 nm
R_{max}	$\tanh^2(\kappa L)$	—
T_{min}	$\cosh^{-2}(\kappa L)$	$\cos^2(\kappa L)$

close is the difference $\Delta\beta$ between the propagation constants of the coupled modes to the spatial frequency $2\pi/\Lambda$ of the periodic perturbation which constitutes the grating. Maximum coupling is for $\delta = 0$ which yields the resonance wavelengths λ_{max} which is approximately equal to the design wavelength λ_D equalling the FBG or LPG wavelengths λ_B or λ_{LPG}.

Table 2 summarizes the basic parameters: the detuning parameter δ, the resonance wavelength λ_{max}, the spectral bandwidth $\Delta\lambda_0$ between adjacent zeroes, the power transmission T and power reflection R coefficients. In the above expressions, the α_{BRF} is the bandwidth refinement factor. For a FBG two extreme sub-cases for the grating bandwidth $\Delta\lambda_0$ are of importance:

- strong FBGs for which $\Delta nL >> \lambda_B$ so $\Delta\lambda_0/\lambda \propto \Delta n_0$.
- weak FBGs for which $\Delta nL << \lambda_B$ so $\Delta\lambda_0/\lambda \propto 1/L$.

Since uniform FBGs have strong side lobes apodized gratings are preferred. Typical reflection and transmission spectra of an apodized FBG are presented in Fig. 4.

For apodized gratings the side lobes are stongly suppressed, so depending on the FBG strength, the bandwidth is given at specific levels such as the 0.1 dB, the 3 dB or for example the 10 dB. For LPGs, $\lambda_{m,LPG}$ is a separate resonance wavelength corresponding to each higher order cladding mode. The higher the mode order m, the greater the difference $n_0 - n_m$ and hence, the higher the resonance wavelength $\lambda_{m,LPG}$ as is shown in Fig. 5.

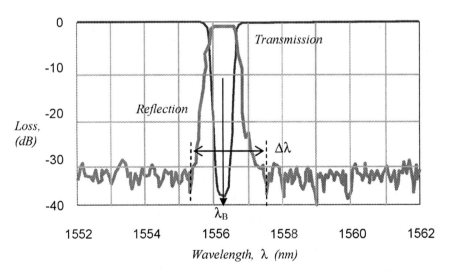

Figure 4. Reflection and transmission spectra of a strong apodized FBG.

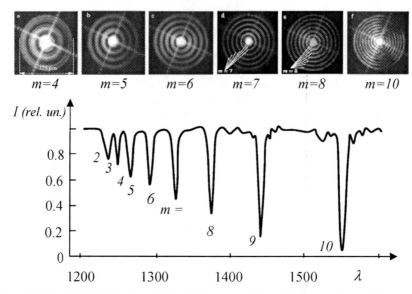

Figure 5. Near field distributions of the higher order HE_{1m} cladding modes and the corresponding spectral response of the LPG (Vasiliev, 1999).

3. FBG and LPG Fabrication Techniques

When considering FBG and LPG fabrication techniques we must first note that FBG pitch length is $\Lambda_{FBG} \approx 0.5$ μm while $\Lambda_{LPG} \propto 10^3 \, \Lambda_{FBG}$. The scales being so different, we have a limitation to the FBG fabrication method and a greater freedom to produce LPG index modulation.

3.1. FBG FABRICATION TECHNIQUES

3.1.1. *Photosensitivity*

All FBG fabrication techniques are based on the photosensitivity pheno-menon which is observed as a change of the refractive index of silica glass when irradiated with high energy UV photons. For the purposes of sensor applications two important things must be outlined: the specific methods to achieve maximum sensitivity and the effect on the optical and mechanical properties of the gratings.

To achieve maximum sensitivity several approaches are used. First, choice of an appropriate wavelength of maximum absorption. Typically these are around 330 nm, 242 nm, 193 nm and 157 nm. The lasers used are then 488 nm and 514 nm Ar-ion, 248 nm KrF-excimer, frequency-doubled 244 nm and 257 nm Ar-ion, frequency doubled XeCl-pumped dye, 197 nm ArF-excimer (Albert, 1994) and 157 nm F_2 (Herman, 1997). Second, addition of dopants as Ge (Hill, 1978; Meltz, 1989), with B- (Williams, 1993) or Sn-codopings (Dong, 1995) as well as P (Strasser, 1995) and N (Dianov, 1997) to increase photosensitivity at 193 nm. Third, H_2 loading of the fibre usually under high pressure and low temperature usually increases photosensitivity by an order of magnitude (Lemaire, 1993).

UV-induced refractive index changes are accomapnied by two major effects: mechanical weakening and optical instability. The mechanical strength of a FBG compared to the pristine fiber is reduced up to 4 times (James, 1999) since the UV radiation causes disbonding and defects in the glass network. This entails reliability issues for strain sensor applications. Optical instability arises because the photoinduced defects are unstable and with time the fiber relaxes to its initial state which is equivalent to erasing the FBG. Complete erasure at normal ambient conditions may take hundreds of years but the process is initially extremely fast causing instability of the Bragg wavelength and the bandwidth. Accelerated optical aging must then be used to stabilize the FBG (Erdogan, 1994).

3.1.2. *UV Laser Writing Techniques*

There are several techniques for UV inscription in fiber gratings, all based on interference, which allow to achieve the small values Λ_{FBG}.

The first to be discovered were the self-induced Bragg gratings (Hill, 1978) which were in fact imprintings of a standing wave along an optical fibre. These gratings have no further applications and development.

The interferometric methods (Meltz, 1989) require coherent UV lasers (L_{coh} = 2 cm to 10 cm) (Ar, dye) and a stable interference pattern. The Talbot (Bennion,1996) and Lloyd interferometer arrangements allow flexibility of FBG parameters since Λ, and hence λ_B, can be varied.

The phase mask method (Hill, 1993) is largely preferable for mass-production since it ensures repeatability of parameters and makes use of cheaper, less coherent excimer UV lasers. The methods suffers from a limited flexibility but it can be increased using an additional Talbot (Dyer, 1995) interferometer.

Writing of FBGs during the drawing process is a third approach which allows multiple in-line FBGs to be written in series. It is appropriate for time division multiplexing (TDM) FBG sensor systems which use identical gratings spaced at equal distances. As the fiber is not stripped the strength of the grating is higher. A drawback of the method is its costliness.

3.2. LPG FABRICATION TECHNIQUES

The fact that the pitch length of the LPG is hundreds of microns allows simpler and lower cost approaches to be used.

3.2.1. *UV Laser Writing Techniques*

The use of amplitude phase masks with UV irradiation allows the whole LPG to be written simultaneously. The point-by-point method (James, 2003) allows for individual fine tuning of the LPG spectrum. The advantages of the UV-inscription are good repeatability, low polarization dependent loss (PDL) but the drawbacks are a weakening of the grating and the need for post-inscription thermal annealing. Also, the spectrum of the grating will be very much dependent on the type of laser used, the type of fibre and the use of hydrogen loading (Chen, 2001).

3.2.2. *High Temperature Heating: CO_2/CO-laser Writing and Arc- discharge Techniques*

Periodic index modulation can be achieved using localized thermal treatment (Dianov, 1997) of the fiber by heating to temperatures up to 1000°C. There are two basic ways to achieve that: either CO_2 laser (Davis, 1998) and CO laser heating (Dianov, 1997), or electric arc-discharge (Godbout, 1998; Kosinski, 1998; Rego, 2005). There are several possible mechanisms that lead to the change of the refractive index as a result of thermal action: mechanical deformation of the fibre structure creating smooth tapers (Kosinski, 1998); redistribution of previously frozen elastic stresses (Enomoto, 1998); diffusion of the core dopants (Dianov, 1997; Palai, 2001); frozen-in viscoelsticity (Yablon, 2004); modifications of the glass structure (Rego, 2001; Malki, 2003).

The CO_2/CO-laser and arc-discharge techniques have several obvious advantages: the technology is cheaper and flexible since it is a point-by-point writing process; it is fast and is applicable to any type of fibre – silica, photonic crystal fibers (PCF) and microstructured fibers. It has also several disadvantages, the first being its limitation to pitch lengths $\Lambda > 200$ μm, higher PDL and poor repeatability especially for arc-induced LPGs. PDL is highly undesirable for LPG sensors since the resonance wavelength and the response of the grating will be strongly is polarization-dependent. PDL of several dBs has been observed in CO_2-written (Zhu, 2000; Bachim, 2003) and up to 9 dB has been reported in arc-induced (Rego, 2006) LPGs. To overcome PDL azimuthally symmetric exposures (PDL ≤ 0.21 dB), fiber rotation (PDL ≤ 0.25 dB) and a helical structure (PDL ≤ 0.42 dB) of CO_2-written LPGs have been suggested (Ishii, 2002; Oh, 2004). Also, fabricating LPGs with a KrF laser reduces PDL to 0.25 dB (Oh, 2003).

3.2.3. *Mechanically Induced and Corrugated LPGs*

Mechanically induced LPGs have been demonstrated in silica (Lin, 2004), PCF (Lim, 2004) and plastic (Hiscocks, 2006) optical fibers and while this technique is easy to implement, it, too suffers form PDL. It also remains unclear as to the general reliability of such structrues subjected to a constant stress.

4. Fiber Bragg Grating Sensor Applications

4.1. FBG SENSORS

FBGs have appeared as attractive for sensing applications for three very simple reasons:

- the first is that the center wavelength λ_B shifts linearly with temperature and strain which are thus spectrally encoded;
- the second is that changes of light intensity are unimportant;
- the third is the large possibility for sensor multiplexing and networking.

4.1.1. Sensitivities

Any sensor application is dependent on sensitivities to external physical fields. The relative center wavelength shift of silica fiber based Bragg gratings can be written as:

$$\delta\lambda/\lambda_B = K_\varepsilon\,\delta\varepsilon + K_T\,\delta T + K_p\,\delta p \qquad (1)$$

where K_ε, K_T and K_p are the strain, temperature (Kersey, 1997) and pressure (Xu, 1993) coefficients.

The corresponding sensitivities $S_\varepsilon = \lambda_B K_\varepsilon$, $S_T = \lambda_B K_T$ and $S_p = \lambda_B K_p$ at three different wavelengths are given in Table 3.

There is also a very weak dependence on magnetic field which for the usual ambient conditions is negligible. FBGs are practically insensitive to bending, twisting and ambient refractive index changes.

As the minimum detectable wavelength shift is of the order of 1 *pm*, only temperature and strain sensing is then of practical importance. Therefore most of the applications for a FBG sensor is strain, or strain-related sensing. In the latter case a transducer to strain is needed.

TABLE 3. Fiber Bragg grating sensitivity to strain (S_ε), temperature (S_T) and pressure (S_p).

λ	S_ε (pm/$\mu\varepsilon$)	S_T (pm/°C)	S (pm/bar)
850 nm	≈ 0.6	7 – 7.84	–
1300 nm	1.051	10.5	–
1550 nm	1.254	12.5	≈ 0.43

4.1.2. *Single Parameter Sensors*: *Strain-based and Temperature FBG Sensors*

To have a truly single-parameter FBG sensor, the sensitivity to the other parameter has to be suppressed or compensated. For a strain sensor temperature dependence has to be elliminated. Using appropriate transducers, strain-based sensors can measure mechanical stress, load, current, flow, inclination, angle of rotation, relative humidity etc. A number of practical implementations can be enumerated to illustrate the large variety of solutions.

Mechanical stress sensing is perhaps the largest application of FBGs. The grating is mounted in a package which is attached to different constructions to measure strain. The techology is mature and a number of cumpanies offer commercial products for long term health monitoring of civil structures, crack detection in dams or bridges, vibration monitoring of bridges and masts etc.

Load/pressure sensing is an option of mechanical stress sensing and a matter of an appropriate package to tranduce load into longitudinal stress as for example a bimetal thermally compensated FBG load sensor (Tian, 2005). The basic application is load watch of bridges or platforms and strructural health monitoring (SHM). A pressure transducer with low temperature sensitivity has been proposed for the oil and gas industry (Yamate, 2002).

Electrical current sensing or electromagnetic force sensing is possible with an appropriate current-to-strain transducer using differential signal from two FBGs on opposite sides of a cantilever (Zhao, 2005, 2006). Flow sensing with a cantilever has also been proposed (Zhao, 2005). Inclination sensing has also been demonstrated using the cantilever transducer (Zhao, 2005). Relative humidity FBG sensor using moisture sensitive polymer coating inducing strain through volume expansion has also been tested (Yeo, 2005). Water level sensing has been proposed (Fukuchi, 2002) on the basis of a mechanical level-strain transducer.

A FBG hydrophone with flat frequency response (Takahashi, 1999) and a underwater acoustical sensing array with temperature compensation (Tanaka, 2006) have been demonstrated successfully.

4.1.3. *Two-parameter Sensing*

The alternative to single-parameter sensing is two-parameter sensing in which both the strain and the temperature are measured. In this case sensitivity and cross-sensitivity coefficients must be known and a system of two linear equations solved (Othonos, 1999).

$$\begin{pmatrix} \varphi_1 \\ \varphi_2 \end{pmatrix} = \begin{bmatrix} K_{1T} & K_{1\varepsilon} \\ K_{2T} & K_{2\varepsilon} \end{bmatrix} \begin{pmatrix} T \\ \varepsilon \end{pmatrix} \qquad |\delta\varepsilon| = \frac{|K_{2T}||\delta\varphi_1| + |K_{1T}||\delta\varphi_2|}{|K_{1T}K_{2\varepsilon} - K_{2T}K_{1\varepsilon}|} \qquad (2)$$

In this case the FBG hase to be operated at two different wavlengths or it should be written in a birefringent FBG.

4.2. FBG MULTIPLEXING AND INTERROGATION TECHNIQUES

4.2.1. *Multiplexing Techniques*

There are three major multiplexing techniques for FBGs listed in Table 4.

TABLE 4. Comparison of the multiplexing techniques

Features	TDM	WDM	SDM	SDM/WDM
Multiplexing capacity	medium	good (with TDM)	good	very good
Spatial resolution	low	high	high	high
Usage of optical power	good	good	good (with switching)	good (with switching)
Inter-changeability	low	low	high	medium
Potential cost	low	medium	medium	medium

Time division multiplexing (TDM) needs weak FBGs with identical center wavelengths spaced at equal distances along the fiber. They can be written during fiber drawing. Pulses are reflected and the signal from each grating is identified by the time delay it suffers. Wavelength division multiplexing (WDM) can make use of strong gratings placed at any distance along the fiber. Demultiplexers are then used to separate the signal from each grating.

Spatial divison multiplexing (SDM) is based on a parallel topology and is best suited when single replaceable and interchangeable FBGs have to be individually addressed. The comparison Table 4 (Othonos, 1999) gives a good comparison between the tree types of gratings.

4.2.2 *Interrogation Techniques*

FBG interrogation techniques must ensure a resolution of $\delta\lambda_B \approx 1\ pm$ which translates into temperature and strain resolutions of $\delta T \approx 0.1°C$ and $\delta\varepsilon \approx 1\ \mu\varepsilon$, correspondingly. Several methods have been suggested and implemented in practical devices.

The use of an edge filter is the simplest technique (Melle, 1992) and requires a narrow bandwidth FBG ($\Delta\lambda \approx 0.05$ to 0.3 nm) whose wavelength shifts of the reflected signal are tranformed into intensity changes by a linearly varying filter. It is either a low-pass filter or a low-pass edge for greater sensitivity. A splitter provides a reference signal I_R and the signal to reference ratio $I_S/I_R = A(\lambda_B - \lambda_0 - b)$ linearly depends on the shift of λ_B. A WDM fiber coupler (Davies, 1994) or a wide chirped FBG can be used for an all-fiber version.

The use of tunable filters is a way to increase the dynamic range. Proposed options are a tunable wavelength fiber Fabry-Perot filter (Kersey, 1993), tunable FBG filters (Jackson, 1993), matched FBG filters for FBG arrays (Davies, 1995) and acousto-optic tunable filters (Xu, 1996). Tunable filters are widely used, but they have a serious energetic disadvantage. If the FBG reflects an energy E_g, $\delta\lambda_f$ is the filter bandwidth and $\Delta\lambda_s$ is the width of the scanned wavelength range, then the detected amount of the energy $E_D = E_g . \delta\lambda_f/\Delta\lambda_s$. Thus most ($\approx 99\%$) of the reflected energy is lost, which in turn demands strong gratings and high power sources.

Interferometric (dynamic sensing) is of prime importance when high resolution and a dynamic range as high as 100 dB are needed and slowly varying changes are not of interest. Intereferometric schemes convert wavelegth shifts into phase shifts. Usually asymmetric unballanced Mach-Zender (Kersey, 1992) and Michelson interferometers are used.

An alternative to tunable filters is the use of a single-frequency, narrow-linewidth wavelength tunable source such as an Er fiber laser (Ball, 1994). The advantage is the improved signal-to-noise ratio. The disadvantage is the limited tuning range of the source typically lass than 2.5 nm.

A CCD-based interrogation system needs a bulk diffractioin grating to disperse the reflected spectrum from the FBGs which is incident upon the pixels of the CCD array. Since typically a pixel-to-pixel spacing corresponds to 100 pm, a weighted average of the illuminated pixels is needed to achieve a 1 pm resolution. The advantages of the method is that using a two-dimensional CCD array, both spatial and wavelength division demultiplexing can be achieved (Chen, 1997) and, unlike tubable filters, all optical power is efficiently used.

Table 5 presents a short comparison of the enumerated interrogation techniques (Othonos, 1999).

TABLE 5. Comparison of the different interrogation techniques

Characteristics	Edge filter	Tunable filter	Inter-ferometric	Tunable laser	CCD spectrometer
Range to resolution	$10^2 - 10^3$	$10^3 - 10^4$	$10^3 - 10^4$	$10^3 - 10^5$	$10^3 - 10^4$
Measurement speed	high	high	high	high	high
Long-term stability	good [1]	good [2]	good	good	good
WDM compatibility	low	high	high	high	high
Potential cost	low	medium	medium	high	low

(1) requires filter stabilisation; (2) requires reference grating

4.3. FBG SENSORS AND NETWORKS

The need for sensor systems for structural health monitoring (SHM) i.e. the assessment of strain/stress, impact and damage with both civil and military applications is one of the main driving forces for the intensive development of multiplexed FBG sensor networks. The FBG sensor network may be installed in either one large constuction as bridges, ships, airplanes, or in just certain critical parts of the system to provide a real-time monitoring. Different multiplexing and interrogation schemes are used depending on the particular application, topology and requirements of the system.

4.3.1. *Civil Structures*

Applications in civil structures (Schulz, 1998; Moyo, 1998) such as bridges, highways, high-rise buildings, mines may be targetting different stages of the construction: during-construction and after-construction stress monitoring (Vohra, 1999).

The during-construction measurements includes: (i) pre-stress forces applied to pre-cast concrete components; (ii) strain at critical locations; (iii) disbonds between the concrete and the metal grid; (iv) concrete shrinkage and chemistry (pH) during hardening.

The post-construction measurements include: (i) continuous detection of incidents and impacts; (ii) evaluation of aging and micro-cracking of concrete and asphalt; (iii) weigh-in-motion (WIM).

Both static and dynamic measurements are taken from embedded or surface mounted FBG. The SHM systems reported (Vohra, 1999) use spectral multiplexing, Fabry-Perrot filters and up to 64 surface-mounted or embedded FBGs. FBG sensor systems within six projects undertaken by ISIS Canada have been reported (Tennyson, 2001) to measure static and dynamic loads on bridge decks and columns, including composite repairs for rehabilitation purposes. Gages with lengths from 1–20 m have been developed for bonding to the concrete structure or embedding in the composite repair patches. System reliability over several years in a hostile environment has been studied.

4.3.2. *Aerospace and Avionics*

FBG sensor systems have become highly attractive for SHM in both civil and militaty aircrafts. A system of 8 strain and 6 temperature sensors in the fuselage of Airbus A 340–600 (Betz, 2002) was successfully tested in parallel with electrical strain gages to measure load in real conditions.

The European Space Agendy plans to use fiber optic sensors and FBGs in particular in future space structures (McKenzie, 2005). An interesting application of chirped FBGs, which are longer (50 mm) and with a broader spectrum ($\Delta\lambda \sim 5nm$), is the detection of disbonding of laminates and crack formation which introduce changes in the spectral distribution.

4.3.3. *Marine*

The application of FBG sensor networks for SHM in marine vessels has been known for ten years (Kersey, 1997). An example of a successful test is the 44 FBG sensor system (Sagvolden, 2002), mounted on a Royal Norwegian Navy Mine countermeasure vessel for ship hull monitoring. The network uses a tunable Fabry Perrot filter interrogation system and a broadband source. Similar systems have been installed on two oil tankers.

4.3.4. *Oil and Gas Industry Applications*

An important and challenging application of FBG sensors and networks is in the oil industry with two main areas of interest (Eigenraam, 1999).

The first is real-time data from the reservoir which means measurement of downhole pressure, temperature and flow. Basically, point sensors are needed in this case for a downhole pressure up to 1000 bars and downhole temperatures up to 250°C. The second area of interest is field applications in a refinery and monitoring of pipe lines where a number of points and processes have to be monitored so a sensor network is needed.

5. Long Period Grating Sensor Applications

5.1. LPG SENSORS

5.1.1. *LPG Sensitivities*

Unlike FBGs in which counterdirerctional coupling occurs in the core, codirectional coupling in LPGs is between a core and a cladding mode, and this fundamental physical difference has the following major effects.

First, the LPGs are sensitive not only to temperature T and strain ε, but also to bending causing a curvature C, to hydrostatic pressure p, to torsion τ and to ambient refractive index n changes.

The relative center wavelength shift is then written as:

$$\delta\lambda / \lambda_{LPG} = K_T \delta T + K_\varepsilon \delta\varepsilon + K_p \delta p + K_C \delta C + K_\tau \delta\tau + K_n \delta n \tag{3}$$

Second, as the center wavelength λ_{LPG}^m is mode-order dependent, so are in the general case the coefficients $K_i = K_i^m$ ($i = T, \varepsilon, C, p, \tau, n$) which can also change sign.

Third, the coefficients K_i^m are fiber- and technology-dependent, because as outlined earlier the cladding mode propagation constants vary with fiber type, with doping contents or with H_2 loading.

Two important consequences arise from the above three effects. The good consequence is that we have far greater possibilities for sensing a variety of measurands, a greater variety of sensitivities and a larger opportunity for multiparameter sensing (Bhatia, 1999). The bad consequence is that from technological point of view a far greater care must be taken to ensure repeatability of the LPG sensitivity parameters and when a single paramerter is to be measured the rest of the parameters need to be suppressed. This may cause a serious problem. For example, recoating of the LPG for the sake of physical protection will change all of the resonnances λ_{LPG}^m and possibly the sensitivities compared to a uncoated LPG. Also, thermal compensation at one resonance wavelength changes the sensitivities at the other wavelengths, but does not suppress their thermal dependence.

Temperature sensitivities S_T have been found (Bhatia, 1999) to vary from 38 pm/°C (1159.4 nm) to 40.48 pm/°C (1219.7 nm), 48.3 pm/°C (1332.9 nm), and 100 pm/°C (1608.6 nm) for H_2 loaded SMF-28 fiber-based UV-written LPG. However, for an LPG written with femtosecond 248 nm UV pulses in a H_2-free and H_2-loaded *Fibercore PS1250/1500* fiber, $S_T = -40$ pm/°C (1464 nm), while in a H_2-loaded *Nufern GF1* fiber, $S_T = -6.7$ pm/°C (Kalachev, 2005). These examples show that S_T varies from -6.7 pm/°C to $+100$ pm/°C versus ≈ 12 pm/°C for SMF-28-based FBGs.

Temperature sensitivity is reduced at maximum for PCF-based LPGs. PCFs are made of a single material so the differential expansion of core and cladding in standard silica fibers is absent in their case. It has been found (Bock, 2006) that for PCF-based tapered arc-induced LPGs, $S_T \approx 0.35$ pm/°C at 1550 nm, which is comparable to FBG sensitivity in a thermally compensated package.

Strain sensitivity, too, is mode order dependent. For the H_2-loaded SMF-28 fiber-based UV-written LPG, the sensitivities S_ε have been found (Bhatia, 1999) to be -0.013 pm/$\mu\varepsilon$ (1159.4 nm), 0.071 pm/$\mu\varepsilon$ (1219.7 nm), 0.29 pm/$\mu\varepsilon$ (1332.9 nm) and 3 pm/$\mu\varepsilon$ (1608.6 nm) vs. 1.16 pm/$\mu\varepsilon$ for a FBG. In a PCF-based tapered LPG, $S_\varepsilon = -2.76$ pm/$\mu\varepsilon$ has been reported (Bock, 2006) in a reflective configuration.

Pressure sensitivity has been mainly studied in tapered LPGs. In SMF-28 fiber based LPGs working in a reflective mode $S_p = 5.1$ pm/bar with $S_T = 49.5$ pm/°C (Bock, 2006) at 1580 nm. With the same arrangement in a PCF-based LPGs $S_p = 11.2$ pm/bar but $S_T \approx 0.35$ pm/°C (Bock, 2006).

TABLE 6. Dependence of the sensitivity $S_C = \delta\lambda/\delta C$ on the type of LPG

Type of LPG	S_C (nm.m)	Reference
UV-written LPG	49.3	(James, 2000)
Phase shifted LPG	31	(Han, 2002)
Pair of LPGs	14.4	(Han, 2000)

Bending of the LPG causes a variation in the periodicity, a tilt in the period and a strain-induced $\Delta\lambda$ (Arce-Diego, 2000) change as well as a split in the spectral response (Shu, 1999). Changes in the curvatrure $C = 1/R$, where R is the bending radius, lead to wavelength changes and the sensitivity $S_C = \delta\lambda/\delta C$ depends on the types of LPGs under test as Table 6 summarizes.

Sensitivity to torsion S_τ, given as the wavelength shift per twist rate, depends on the type of LPG under test. Table 7 is a brief summary of some results.

TABLE 7. Dependence of the sensitivity to torsion S_τ on the type of LPG

Type of LPG	S_τ (pm/rad/m)	Range (rad/m)	Reference
UV-written	49.3	≈ 220	(Gonzalez, 2004)
CO_2-written	74.28	≈ 17.5	(Wang, 2002)
	224	40	(Rao, 2006)
Arc-induced	637	12.6	(Inn, 2002)
Corrugated	1333	<30	(Lin, 2001)

We see a dramatic difference in the sensitivities and the range of operation depending on the LPG fabrication technology. Sensitivity to refractive index n changes unlike those to temperature, strain, torsion and bending, are highly nonlinear. The closer the ambient refractive to that of the cladding and the higher the order of the cladding mode, the stronger the S_n sensitivity. Thus, at 1608.6 nm, very weak sensitivity $S_n = \delta\lambda/\delta n \approx 0.084$ pm/10^{-6} is observed for $n = 1.32$ to 1.4 and an extremely strong sensitivity $S_n \approx 8.75$ pm/10^{-6} for $n = 1.44$ to 1.444 (Bhatia, 1999). It is this high sensitivity that causes an enormous interest in the development of various types of refracive index-based LPG sensors.

5.1.2. LPG Sensor Applications

Temperature sensing in nuclear environments has been shown to be very promising (Fernandez, 2004) with arc-induced LPGs ($S_T = 29$ pm/°C) due to their insensitivity to γ-radiation for levels as high as 493 kGy. As commented, in contrast to FBGs, LPGs show promising future applications for bending and twist measurements in civil structures. An embedded bending sensor has recently been demonstrated (Tan, 2006). Studies are, however, in an early phase compared to FBGs.

Refractive index measurement enjoys the greatest attention so far due to the already mentioned high sensitivities. Sensing of aromatic organic compounds detecting concentrations as low as 0.04% has been demonstrated (Allsop, 2001) using LPGs. Ionic self-assembled multilayers (ISAM) have been shown (Wang, 2005) to be able to fine-tune the resonant wavelengths by varying the thickness of the multilayer. Since various particles can be incorporated into the nanometer thick multilayers, they become promising for biosensing applications.

Concentration of cane sugar was measured with 1300 nm and 1550 nm LEDs using a 248 nm UV written LPG (Shu, 1999). The dual peak wavelength separation was measured in this sensor. A chemical sensor based on a LPG in a multimode fiber was reported to measure chamical concentrations as low as 10 nmol/l (Lee, 2003). Concentration of sugar, salt and ethylene glycol have been measured with a tapered LPG refractometer (Chong, 2004). Measurement of sucrose concentrtions using self-assembnled gold colloid upon a CO_2-written LPG has been suggested (Tang, 2006) and both the LPG wavelength and depth changes have been measured.

Cascaded LPGs forming in-line Mach-Zender or in-line Michelson interferometers offer a great potential for highly senisitive fiber optic refractometers or liquid-level meters (James, 2003).

5.2. LPG MULTIPLEXING

From the point of view of multiplexing with LPGs we have again fundamental differences compared to FBGs. In FBGs we can practically apply dense wavelength division multiplexing (DWDM) because the gratings are narrow ($\Delta\lambda \leq 0.3$ nm) and a sensing channel would occupy < 3 nm of bandwidth. With LPGs, $\Delta\lambda_{LPG} \approx 10$–40 nm, so DWDM is impossible. Instead, we have the possibility to use more than one resonnance wavelength which would then mean a coarse WDM (CWDM) but the sensitivities may be vastly different.

6. FBG and LPG Sensor Comparison

Table 8 summarizes the basic differences in the progress of development of FBG and LPG sensor systems.

The exciting applications of FBG sensors are large multiplexed sensor networks for SHM in a variety of fields. The exciting applications for LPGs at the moment are a limited number of specific sensors mostly in areas alternative to those optimal for FBGs such as bio- and chemical sensors based on refractive index measurements using overlayed structures as well as bending, tosion and pressure sensing.

TABLE 8. Comparison between FBG and LPG sensors

Characteristics	FBGs features	LPGs features
Fabrication technology	Mature	Research going on
Packaging technology	Mature	Non-developed
Reliability	Data accumulated	More data needed
Multiplexed sensors	Large number	Small number
Networking capabilities	Excellent	Limited
Measurable parameters	T and ε	T, ε, p, C, τ and n
Multiparameter sensing	Two-parameter	Multiparameter

References

Albert, J., Malo, B., Bilodeau, F., Johnson, D., Hill, K., Hibino, Y. and Kawachi M. (1994) Photosensitivity in Ge-doped silica optical waveguides with 193 nm light from an ArF excimer laser, *Opt. Lett.* **19**, 387–389.

Allsop, T., Zhang, L. and Bennion I. (2001) Detection of aromatic organic compounds in paraffin by a long-period fiber grating optical sensor with optimized sensitivity, *Optics Commun.* **191**, 181–190.

Arce-Diego, J., Gonzalez-Fernandez, D., Quintela, M., Madruga, F., Lopez-Higuera, J. (2000), Spectral charactersitics of curved long-period gratings, *OFS-14*, Venice, Italy, P2-43, 850–853.

Bachim, B. and Taylor, T. (2003) Polarization-dependent loss and birefringence in long-period fiber gratings, *Appl. Opt.* **42**, 6816–6823.

Ball, G., Morey, W. and Cheo, P. (1994) Fiber laser source/analyzer for Bragg grating sensor array interrogation, *J. Lightwave Techn.*, **12**, 700–703.

Bennion, I., Williams, J. and Zhing L. (1996) UV-written in-fiber Bragg gratings, *Opt. and Quant Electron.* **28**, 93–135

Betz, D., Staudigel, L. and Trutzel, M. (2002) Test of a Fiber Bragg Grating Sensor Network for Commercial Aircraft Structures, *OFS-15*, TuA2, 55–58.

Bhatia, V. (1999) Applications of long-period gratings to single and multi-parameter sensing *Optics Express*, **4**, 457–466.

Bock, W., Chen, J., Mikulic, P. and Eftimov, T. (2006) A Novel Fiber-Optic Tapered Long-Period Bragg Grating for Pressure Monitoring, *Instrumentation and Measurement Technology Conference, IMTC-2006*, Sorento, Italy, 1942–1945.

Bock, W.J., Chen, J., Mikulic, P., Eftimov, T. and Korwin-Pawlowski, M. (2006) Pressure Sensing Using Long-Period Tapered Gratings Written in Photonic Crystal Fibers, *OFS-18*, Cancun, Mexico, Th A6

Chan, T., Yu, L., Tam, H., Ni, Y., Liu, S., Chung, W. and Cheng, L. (2006) Fiber Bragg grating sensors for structural health monitoring of Tsing Ma bridge: Background and experimental observation, *Engineering structures* **28**, 648–659.

Chen, S. (1997) Digital spatial and wavelength domain multiplexing of fiber Bragg grating based sensors, *Proc. of Optical Fiber Sensors Conference, OFS-12*, Williamsburg, VA, USA, 456–459.

Chen, K., Herman P., Zhang, J. and Tam, R. (2001), Fabrication of strong long-period gratings in hydrogen-free fibers with 157-nm F_2-laser radiation, *Opt. Letts*. **26**, 771–773.

Chong, J., Shum, P., Haryono, H., Yohana, A., Rao, M., Lu, C. and Zhu, Y. (2004) Measurement of refractive index sensitivity using a long-period grating refractometer, *Optics Commun*. **229**, 65–69.

Davies, M. and Kersey, A. (1994) All-fiber Bragg grating strain sensor demodulation technique using a wavelength division coupler, *Electron. Lett*. **30**, 75–77.

Davies, M. and Kersey, A. (1995) Matched filter interrogation technique for FBG arrays, *Electron, Lett*. **31**, 822–823.

Davis, D., Gaylord, T., Glytis, E., Mettler, S. (1998) CO_2 laser-induced long-period fibre gratings: spectral characteristics, cladding modes and polarisation independence, *Electron. Lett*. **34**, 1416–1418.

Dianov, E., Golant, K., Krpatko, R., Kurkov A., Leconte, B., Douay, M., Bernage, P., Niay, P. (1997) Strong Bragg grating formation in germanium free nitrogen doped fibers, *OFC'97*, PD5

Dianov, E., Karpov, V., Grekov, M., Golant, K., Vasiliev, S., Medvedkov, O., Khrpatko, R. (1997) Thermo-induced long period fibre gratings, *Proc. Europ. Conf. On Opt. Commun*. Edinburgh, UK, 53–55.

Dong, L., Cruz, L., Reekie, L., Xu, M. and Payne D. (1995) Large photo-induced index change in Sn-codoped germanosilicate fibers, *OSA Techn. Dig. Series* **22**, SuA2, 70–73.

Dyer, P., Farley R. and Giedl, R. (1996) Analysis and application of a 0/1 order Talbot interferometer for 193 nm laser grating formation, *Optics Comm*. **129**, 98–108.

Eigenraam, P., Douma, B. and Koopman, A. (1999) Applications of Fiber Optic Sensors and Instrumenttion in the Oil and Gas Industry, *OFS-13*, F2-1, 602–607.

Enomoto, T., Shigehara, M., Ishikawa S., Danzuka, T. and Kanamori, H. (1998) Kong-period fiber grating in pure-silica core fiber written by residual stress relaxation, *Proc. Opt. Fiber Commun. Conf*., Washington DC, OSA, ThG2, 277–278.

Erdogan, T. (1997) Fiber grating spectra, *J. of Lightwave Techn*., **15**, 1277–1294

Erdogan, T., Mizrahi, V., Lemaire, P. and Monroe, D. (1994) Decay of ultraviolet-induced Fiber Bragg gratings, *J. Appl. Phys*, **76**, 73–80.

Fernandez, A., Gusarov, A., Brichard, B., Bodart, S., Lammens, K., Berghmans, F., Decreton, M., Megret, P., Blondel, M. and Delchambre, A. (2002) Temperature monitoring of nuclear reactor cores with multiplexed Fiber Bragg grating sensors, *Optical Engineering*, **41**, 1246–1254.

Fernandez, A., Rego G., Gusarov, A., Brichard, B., Santos, J., Salgado, H. and Berghmans, F. (2004) Evaluation of long-period fiber grating temperature sensors in nuclear environments, *Second European Workshop Optical Fiber Sensors, Proc. SPIE*, **5502**, 88–91.

Fukuchi, K., Kojima, S., Hishida, Y. and Ishida Sh. (2002) Optical water level sensors using fiber Bragg grating technology, *Hitachi Cable Review*, **21**, 23–28.

Godbout, N., Daxhelet X., Maurier, A., Lacroix, S. (1998) Long period fiber grating by electrical discharge, *Proc. Europ. Conf. on Opt. Commun*., Madrid, Spain, 397–398

Gonzalez, D. A., Jauregui, C., Quintela, A., Madruga, F. J., Marquez, P., Lopez-Higuera, J.M. (2004) Torsion-induced effects on UV long-period fiber gratings, *Proc. SPIE*, **5502**, 192–195.

Han, Y., Lee, B., Han, W., Paek, U., Chung, Y. (2000) Fiber-optic Bending Sensor with a pir of Long Period Fiber Gratings, P1-21, *OFS-14*, Venice, Italy, 118–121.

Han, Y., Han, W., Paek, U., Chung, Y. (2002) Measurement of bending curvature using band-pass filters based on phase-shifted long-period fibre gratings, TuP5, *OFS-15*, 143–146.

Herman, P., Beckley, K. and Ness, S. (1997) 157-nm photosensitivity in germanosilicate waveguides, *OSA Techn. Dig. Ser.*, **17**, BME4, 159–161.

Hill, K.O., Fujii, Y., Johnson, D.C., Kawasaki, B.S. (1978) Photosensitivity in optical waveguides: Application to reflection filter fabrication, *Appl. Phys. Lett.*, **32**, 647–649.

Hill, K.O., Malo, B., Bilodeau, F., Johnson, D.C. and Albert, J., (1993) Bragg gratings fabricated in monomode photosensitive optical fiber by UV exposure through a phase mask, *Appl. Phys Lett.*, **62**, 1035–1037.

Hiscocks, M., van Eijkelenborg, M., Argyros, A. and Large, M. (2006) Stable imprinting of long-period gratings in microstructured polymer optical fibre, *Opt. Express* **14**, 4644–4649.

In, S., Chung, C. and Lee, H. (2002) The resonance wavelength-tuning characteristic of the arc-induced LPFGs by diameter modulation, *OFS-15*, TuP2, 131–134.

Ishii, I., Shima, K., Okube, S. and Wada, A. (2002) PDL suppression on long-period gratings by azymuthally isotropic exposure, *IEICE Trans. Electron.*, E85-C, 934–939.

Jackson, D., Lobo Ribeiro, A., Reekie, L. and Archambault, J. (1993) Simple multiplexing scheme for a fiber-optic grating sensor network, *Optics Lett*, **18**, 1192–1194.

James, S., Wei, C. and Ye, C. (1999) An Investigation of the Tensile Strength of Fibre Bragg Gratings, *OFS-13*, Kiongju, Korea, 38–41.

James, S., Ye, C., Tatam, R. (2000) Bend Sensing Using Optical Fiber Long Period Gratings, *OFS-14*, Venice, Italy, P1-08, 66–69.

James, S. and Tatam, R. (2003) Optical fibre long period grating sensors: charactersitics and applications, *Meas. Sci. and Techn.*, **14**, R49–R61.

Kalachev, A., Pureur, V. and Nikogosyan D. (2005) Investigation of long-period gratings induced by high-intensity femtosecond UV laser pulses, *Optics Commun.*, **246**, 107–115.

Kersey, A., Berkoff, T. and Morey, W. (1992) High-resolutiion fiber-grating based strain sensor with interferometric wavelength-shift detection, *Electron Lett.* **28**, 236–238.

Kersey, A., Berkoff, T. and Morey, W. (1993) Multiplexed fiber Bragg grating strain sensor system with a fiber Fabry Perot wavelength filter, *Opt. Letts.* **18**, 1370–1372 .

Kersey, A., Davis, M., Patrick, H., LeBlanc M., Koo, K., Askins, C., Putnam, M. and Friebele, E. (1997) Fiber Grating Sensors, *J. Lightwave Techn.*, **15**, 1442–1463.

Kersey, A., Davis, M., Berkoff, T., Dandridge, A., Jones, R., Tsai, T., Cogdell, G., Wang, G., Havsgård, G., Pran, K. and Knudsen, S. (1997) Transient load monitoring on a composite hull ship using distributed fiber optic Bragg grating sensors, *Smart Structures and Materials, Proc. SPIE*, **3042**, 421–430.

Kosinski, S.G. and Vengsarkar, A.M. (1998) Splice-based long period fiber grtings, *Proc. Optical Fiber Communic. Conf.*, Washington, DC, OSA, **2**, ThG3, 278–280.

Lee, B. and Nishii, J. (1998) Self-interference of long-period fibre grating and its application as temperature sensor, *Electron. Lett.* **34**, 2059–2060.

Lee, S., Kumar, R., Kumar, P., Radhakrishnan, P., Vallabhan, C. and Nampoori, V. (2003) Long period gratings in multimode optical fibers: application in chemical sensing, *Optics Commun.*, **224**, 237–241.

Lemaire, P.J., Atkins, R.M., Mizrahi, V., Reed, W.A. (1993) High pressure H_2 loading as a technique for achieving ultrahigh UV photosensitivity and thermaslstability in GeO_2 doped optical fibers, *Electron Lett.*, **29**, 1191–1193.

Lim, H., Lee, K., Kim, J. and Lee, B. (2004) Tunable fiber gratings fabricated in photonic crystal fiber by use of mechanical pressure, *Opt. Lett.*, **29**, 331–333.

Lin, Y. and Wang, L. (2001) A wavelength- and loss-tunable band-rejection filter based on corrugated long-period fiber grating *IEEE Photonics Techn. Lett.,* **13**, 332–334.

Lin, C., Li, Q., Au, A., Jiang, Y., Wu, E. and Lee, H. (2004) Straim-Induced Thermally Tuned Long-Period Fiber Gratings Fabricated on a Periodically Corrugated Substrate, *J. Lightwave Techn.,* **22**, 1818–1827.

Malki, A., Humbert, G., Ouerdane, Y., Boukhenter, A. and Boudrioua, A. (2003), Investigation of the writing mechanisms of electric-arc-induced long-peiod fiber gratings, *Appl. Optics,* **42**, 3776–3779.

Mckenzie, I. and Karfolas, N. (2005) Fiber Optic Sensing in Space Structures: The experience of the European Space Agency, *Proc. SPIE,* **5855**, 262–269.

Melle, S. Liu, K., Measures, R. (1992) A passive wavelength demodulation system for guided-wave Bragg grating sensors, *IEEE Photonics Techn. Lett.* **4**, 516–518.

Meltz,G., Morey W., Glenn, W. (1989) Formation of Bragg gratings in optical fibers by transverse holographic method, *Opt. Lett,* **14**, 823–825.

Moyo, P., Brownjohn, J., Suresh, R. and Tjin, S. (1998) Development of fiber Bragg grating sensors for monitoring of civil infrastructures, *Engineering structures* **27**, 1828–183.

Oh, S., Lee, K., Paek, U. and Chung, Y. (2004) Fabrication of helical long-period fiber gratings by use of a CO_2 laser, *Opt. Lett.* **29**, 1464–1466.

Oh, S., Han, W., Paek, U. and Chung, Y. (2003) Reduction of Birefringence and Polarization-Dependent Loss of Long-Period Fiber Gratings Fabricated with a KrF Excimer Laser, *Opt. Express,* **11**, 3087–3092 .

Othonos, A. and Kalli, K. (1999) Fiber Bragg Gratings, *Artech House*, Boston, London

Palai, P., Saturanayan, M.N., Das. M., Thyagaraja, K. and Pal, B.P. (2001) Characterization and simulation of long period gratings fabricated using electric arc discharges, *Opt. Commun.,* **193**, 181–185.

Rao, Y., Zhu, T. and Mo, Q. (2006) Highly sensitive fibre optic torsion sensor based on an ultra-long-period fiber grating, *Opt. Commun.,* **266**, 187–190.

Rego, G., Okholnikov G., Dianov E. and V. Sulimov (2001) High temperatrure stability of long-period fiber gratings produced using an electric arc, *J. Lightwave Techn.* **19**, 1574–1579.

Rego, G., Marques, P., Santos, J. and Salgado, H. (2005) Arc-Induced Long Period Gratings, *Fiber and Int. Opt.* **24**, 245–259.

Rego, G., Santos, K. and Salgado, E. (2006) Polarization-dependent loss of arc-induced long-period fiber gratings, *Opt. Comm.* **262**, 152–156.

Sagvolden, G., Pran, K., Vines, L., Torkildsen, H. and Wang, G. (2002) Fiber Optic System for Ship Hull Monitoring, *OFS-15*, Portland, Oregon, ThP5, 435–438.

Shu, X., Zhu, X., Wang, J., Jiang, Sh., Shi, W., Huang, Z. and Huang. D. (1999) Dual resonant peaks of LP_{015} cladding mode in long-period gratings, *Electron. Lett.* **35**, 649–651.

Shu, X. and Huang, S. (1999) Highly sensitivie chemical sensor based on the measurement of the separation of dual resonant peaks in a 100-μm period fiber grating, *Optics Commun.* **171** , 65–69.

Schulz, E., Udd, E., Seim, J. and McGill, G. (1998) Advanced fiber-grating strain sensor systems for bridges, structures, and highways*, Proc. SPIE,* **3325**, 212–221.

Strasser, T., White, M., Yan, P., Lemaire, T., Erdogan, T. (1995) Strong Bragg phase gratings in phosphorous doped fiber induced by ArF excimer radiation, *OFC'95*, 159–160.

Takahashi, N., Takahashi, S. and Tetsumura, K. (1999) Fiber Bragg-grating underwater acoustic sensor, *OFS-13*, Kiongju, Korea, Th3-7, 565–568.

Tan, K., Chan, C., Tjin, S. and Dong, X. (2006) Embedded long-period fiber grating bending sensor, *Sensors and Actuators A*, **125**, 267–272.

Tanaka, S., Yokosuka, H. and Takahashi, N. (2006) Temperature independent finer Bragg grating underwater acoustic sensor array using incoherent light, *Acoust. Sci. & Tech.* **27**, 50–52.

Tang, J.L., Cheng, S., Hsu, W., Chiang, T. and Chau, L. (2006) Fiber-optic biochemical sensing with colloidal gold-modified long-period fiber grating, *Sensors and Act. B.* **119**, 105–109.

Tennyson, R., Mufti, A., Rizkalla, S., Tadros, G. and Benmokrane, B. (2001) Structural health monitoring of innovative bridges in Canada with fiber optic sensors, *Smart Mater. Struct.* **10**, 560–573.

Tian, K., Liu, Y. and Wang, Q. (2005) Temperature independent fiber Bragg grating sensor using bimetal cantilever, *Optical Fiber Technology*, **11**, 370–377.

Vasiliev, S.A. (2001) Photoinduced fiber gratings, *Proc. SPIE*, **4357**, 1–12.

Vasiliev, S.A., Dianov, E.M., Medvedkov, O.I., Constantitni, D.M., Iocco, A., Limberger, H.G. and Salathe, R.P. (1999) Properties of the cladding modes of an optical fibre excited by refractive index, *Quantum Electronics* **29**, 65–69.

Vengsarkar, A., Lemaire, P., Judkins, J., Bhatia, V., Erdoga, T. and Sipe J.E. (1996) Long-Period Fiber Gratings as Band-Rejection Filters, *J. of Lightwave Techn.*, **14**, 58–65.

Vohra, S.T., Todd, M.D., Johnson, G.A., Chang, C.C. and Danver, B.A. (1999) Fiber Bragg Grating Sensor System for Civil Structure Monitoring: Applications and Field Tests, *OFS-13*, Tu2-1, 32–37

Wang, Y., Rao, Y., Hu, A., Zheng, X., Ran, Z., Zhu, T. (2002) A novel Fiber-Optic Torsion Sensor Based on a CO_2-laser-induced Long-Period Fiber Grating, *OFS-15*, TuP6, 148–150.

Wang, Zh., Heflin, J., Stolen, R. and Ramachandran S. (2005) Highly sensitivie optical response of optical fiber long period gratings to nanometer-thick ionic self-assembled multipayers, *Applied Physics Letters*, **86**, 223104, 1–3.

Williams, D., Ainslie, B., Armitage, J. and Kashyap, R. (1993) Enhanced UV photosensitivity in Boron codoped germanosilicate fibers, *Electron Lett.*, **29**, 45–47

Xu, M.G. (1993) Optical in-fibre grating high pressure sensor, *Electron. Lett.* **29**, 389–399.

Xu, M., Geiger, H. and Dakin. J. (1996) Modelling and performance analysis of of a fiber Bragg grating interrogation system using an acouso-optic tubable filter, *J. of Lightwave Technol.* **14**, 391–396.

Xu, M., Maaskant, R., Ohn M. and Alavie, T. (1997) Independent tuning of cascaded long period fibre gratings for spectral shaping, *Electron Lett.*, **33**, 1893–1894.

Yablon, D., Yan, M., DiMarchello, F., Fleming, J., Reed, W., Monberg E., DiGiovanni, D., Jasapara, J. and Lines, M. (2004) Refractive index perturbations in optical fibers resulting from frozen-in viscoelasticity, *Appl. Phys. Lett*, **84**, 19–21.

Yamate, T., Ramos, R., Schroeder R., Madhaven, R., Balkunas, S. and Udd, E. (2002) Transversely loaded Bragg grating pressure transducer with mechanically enhancing the sensitivity, *OFS-15*, Portland, Oregon, 525–539.

Yeo, T., Sun, T., Grattan, K., Parry, D., Lade, R. and Powell, B. (2005) Characterisation of a polymer-coated Bragg grating sensor for relative humidity sensing, *Sensor and Actuators B*, **110**, 148–155.

Zhao, Y., Zhao, Y. and Zhao, M. (2005) Novel force sensor based on a couple of fiber Bragg gratings, *Measurement*, **38**, 30–33.

Zhao, Y., Chen, K. and Yang, J. (2005) Novel target type flowmeter based on a differential fiber Bragg grating sensor, *Measurement*, **38**, 230–235.

Zhao, Y., Yang, J., Peng, B. and Yang, S. (2005) Experimental research on a novel fiber-optic cantilever-type inclinometer, *Optics&Laser Techology*, **37**, 555–559.

Zhao, Y., Meng. Q. and Chen, K. (2006) Novel current measurement method based on fiber Bragg grating sensor technology, *Sensors and Actuators A*, **126**, 112–116.

Zhu, Y., Simova, E., Berini, P. and Grover, C. (2000) A Comparison of Wavelength Dependent Loss Measurements in Fiber Gratings, *IEEE Trans. Instrum. Meas.*, **49**, 1231–1239.

NOVEL SENSING MECHANISMS USING TILTED FIBER BRAGG GRATINGS

CHENGKUN CHEN, YANINA Y. SHEVCHENKO
AND JACQUES ALBERT*
*Department of Electronics, Carleton University, 1125 Colonel
By Drive, Ottawa, Canada, K1S 5B6*

Abstract. Fiber Bragg gratings with grating planes tilted at small angles relative to the fiber axis perpendicular couple light to core and cladding modes propagating backwards. The resonant wavelength for each mode depends differentially on external perturbations and thus can be used to make the fiber grating into a multimodal sensor. Using the core mode back reflection resonance as a reference wavelength in single mode fibers, the relative shift of the cladding mode resonances can be used to selectively measure perturbations such as strain or surrounding refractive index, and this independently of temperature. Selected cladding mode resonances allow the measurement of the refractive index of the medium surrounding the fiber for values between 1.25 and 1.44 with an accuracy approaching 1×10^{-4}. Furthermore, when a gold coating with a thickness between 10–50 nm is deposited on the fiber, we show that surface plasmon resonances can be excited at the surface of the gold film, a fiber optic analogy to the Kretschmann plasmon resonance sensor.

Keywords: Fiber sensors, Bragg gratings, refractometry, surface plasmon resonance, strain, sensitivity

1. Introduction

Fiber Bragg grating (FBG) sensors have a wide range of applications such as pressure-strain sensors, temperature sensors, micro-bending sensors and

* Jacques Albert, Department of Electronics, Carleton University, 1125 Colonel By Drive, Ottawa, Canada, K1S 5B6; e-mail: jalbert@doe.carleton.ca

W.J. Bock et al. (eds.), Optical Waveguide Sensing and Imaging, 25–49.

external refractive index sensors[1–6]. As these optical sensors are inherently immune from electromagnetic interference and chemically inert, they are very attractive in bio-chemical applications and hazardous surroundings.

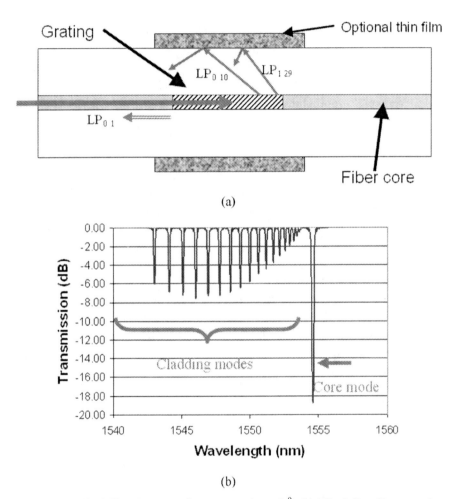

(a)

(b)

Figure 1. (a) Tilted fiber Bragg grating sensor schematic[9], (b) Tilted fiber Bragg grating simulated transmission spectrum.[9]

In this work, we show that a single weakly tilted fiber Bragg grating (TFBG) sensor[7–14] can perform multimodal functionalities, in the same way as a double-FBG sensor or as a long period grating (LPG)-FBG combination. In the TFBG, both a core mode resonance and several cladding mode resonances appear simultaneously shown as Figure 1. This has several advantages. While the cladding mode resonances are sensitive to the parameters of the external environment such as the refractive index, a deposited layer thickness and optical properties, etc. and to physical changes in the whole fiber cross-section (shear strains arising from bending for instance), the core mode (Bragg) resonance is only sensitive to axial strain and temperature. We will show that the temperature dependence of cladding modes is similar to that of the core modes, so that the effect of temperature can be removed from the cladding mode resonance by monitoring the wavelength difference between the core mode resonance and selected cladding mode resonances. Using this technique, temperature-independent strain and surrounding refractive index (SRI) sensors can be produced.

The concept of using cladding mode resonances from a TFBG was originally proposed by Laffont and Ferdinand[8], but their scheme was limited to monitoring the envelope of the resonances as the SRI changes and did not use the core mode resonance for self-referencing with regards to temperature. Our approach makes use of the resonance positions and/or strengths individually, which allows us to extract multiple sensing parameters from a single sensor. To be precise, a multi-functional sensor can be developed by monitoring several cladding mode resonances (or groups of resonances) which have different sensitivities to the external environment changes. For instance, bending affects mainly low order cladding modes[3], while SRI increases affect higher order cladding modes preferentially[8], and neither perturbation affects the core mode resonance.

Furthermore, we will show that a tilted fiber Bragg grating written in standard telecommunications single mode fiber with a 10–50 nm gold coating can excite plasmon resonances on the gold surface. The weakly tilted grating is used to couple the core mode light to a multitude of cladding modes, depending on the light wavelength, as shown in Figure 1a[8,10]. Since the cladding modes have non-zero evanescent fields extending outside the cladding diameter and hence into the metal film, when the axial component of the propagation constant of the cladding mode equals that of an SPR wave, coupling to that SPR wave can occur. In contrast to the LPG and FBG approaches[15–16], a single grating design is sufficient to generate a wavelength dependent set of cladding modes that are "interrogating" the metal film at various angles of incidence. Each of the modes can be individually "addressed" simply by changing the wavelength of the guided light, and each mode strikes the cladding boundary at a different angle of incidence. Therefore several plasmon modes can be excited simultaneously in TFBGs with gold coating, and this novel TFBG sensor can perform as a highly sensitive biochemical sensor through the large local electric field enhancement occurring in the vicinity of the gold surface[15].

2. Analysis

As is well known, the Bragg reflection and cladding mode resonance wavelengths λ_B and λ^i_{clad} of TFBGs are determined by a phase-matching condition and can be expressed as follows[9]:

$$\lambda_B = 2n_{eff} \Lambda / \cos\theta \tag{1}$$

$$\lambda^i_{clad} = \left(n^i_{eff,core} + n^i_{eff,clad} \right) \Lambda / \cos\theta \tag{2}$$

where n_{eff}, $n^i_{eff,core}$ and $n^i_{eff,clad}$ are the effective indices of the core mode at λ_B and the core mode and the ith cladding mode at λ^i_{clad} respectively, and Λ and θ are the period and the internal tilt angle of the TFBG.

2.1. STRAIN PERTURBATIONS OF TFBG

For weakly TFBG, if we consider the Bragg and cladding mode wavelength shifts ($\Delta\lambda_B$, $\Delta\lambda^i_{clad}$) caused by axial strain ($\Delta\varepsilon$) changes only, from equations (1) and (2) the wavelength shifts $\Delta\lambda_B$ and $\Delta\lambda^i_{clad}$ can be written as follows[10]:

$$\Delta\lambda_B = \left(\frac{2n_{eff}}{\cos\theta} \frac{\partial\Lambda}{\partial\varepsilon} + \frac{2\Lambda}{\cos\theta} \frac{\partial n_{eff}}{\partial\varepsilon} \right) \Delta\varepsilon \tag{3}$$

$$\Delta\lambda^i_{clad} = \left(\frac{\left(n^i_{eff,core} + n^i_{eff,clad} \right)}{\cos\theta} \frac{\partial\Lambda}{\partial\varepsilon} + \frac{\Lambda}{\cos\theta} \frac{\partial \left(n^i_{eff,core} + n^i_{eff,clad} \right)}{\partial\varepsilon} \right) \Delta\varepsilon \tag{4}$$

the equations (3) and (4) can be written as follows[17]:

$$\Delta\lambda_B = \lambda_B \left(1 - p_B \right) \Delta\varepsilon \tag{5}$$

$$\Delta\lambda^i_{clad} = \lambda^i_{clad} \left(1 - p^i_{cld} \right) \Delta\varepsilon \tag{6}$$

where

$$p_B = -\frac{1}{n_{eff}} \frac{\partial n_{eff}}{\partial\varepsilon} \tag{7}$$

and

$$p^i_{cld} = -\frac{1}{n^i_{eff,core} + n^i_{eff,clad}} \frac{\partial(n^i_{eff,core} + n^i_{eff,clad})}{\partial\varepsilon} \tag{8}$$

are photoelastic coefficients for core (Bragg) mode and the ith cladding modes respectively, which may be calculated through core and cladding refractive index changes by

$$\delta n = 0.5 * n^3 \left[p_{12} - v(p_{11} - p_{12}) \right] \delta \varepsilon \qquad (9)$$

where p_{11} and p_{12} represent the component of strain-optic tensor, v is Poisson's ratio and n is the core or cladding refractive index. The high order cladding modes have smaller effective indices and hence will have larger photoelastic coefficients than the core mode if the partial derivatives of all the effective indices with respect to a global strain change on the fiber are similar. From equations (5) and (6) and noting that $\lambda_B > \lambda^i_{clad}$ and $p_B < p^i_{cld}$ for high order cladding modes, we obtain:

$$\Delta\lambda_B = \lambda_B \left(1 - p_B\right)\Delta\varepsilon > \lambda^i_{clad}\left(1 - p_B\right)\Delta\varepsilon > \lambda^i_{clad}\left(1 - p^i_{clad}\right)\Delta\varepsilon = \Delta\lambda^i_{clad}$$

So the Bragg peak wavelength should have a larger wavelength shift $\Delta\lambda_B$ than the high order cladding mode resonances wavelength shift $\Delta\lambda^i_{clad}$.

2.2. TEMPERATURE PERTURBATIONS OF TFBG

If we consider the Bragg and cladding mode wavelength shifts ($\Delta\lambda_B$, $\Delta\lambda^i_{clad}$) caused by temperature (ΔT) changes only, from equations (1) and (2) the wavelength shifts $\Delta\lambda_B$ and $\Delta\lambda^i_{clad}$ can be written as follows:

$$\Delta\lambda_B = \left(2\frac{\Lambda}{\cos\theta}\frac{\partial n_{eff}}{\partial T} + 2\frac{n_{eff}}{\cos\theta}\frac{\partial \Lambda}{\partial T} \right)\Delta T \qquad (10)$$

$$\Delta\lambda^i_{clad} = \left(\frac{\Lambda}{\cos\theta}\frac{\partial\left(n^i_{eff,core} + n^i_{eff,clad}\right)}{\partial T} + \frac{n^i_{eff,core} + n^i_{eff,clad}}{\cos\theta}\frac{\partial \Lambda}{\partial T} \right)\Delta T \qquad (11)$$

2.3. SURROUNDING REFRACTIVE INDEX PERTURBATIONS OF TFBG

Consider the Bragg and cladding mode wavelength shifts ($\Delta\lambda_B$, $\Delta\lambda^i_{clad}$) caused by surrounding refractive index (Δn_s) changes only, from equations (1) and (2) the wavelength shifts $\Delta\lambda_B$ and $\Delta\lambda^i_{clad}$ can be written as follows:

$$\Delta\lambda_B = \left(2\frac{\Lambda}{\cos\theta}\frac{\partial n_{eff}}{\partial n_s} + 2\frac{n_{eff}}{\cos\theta}\frac{\partial \Lambda}{\partial n_s} \right)\Delta n_s \qquad (12)$$

$$\Delta\lambda_{clad}^{i} = \left(\frac{\Lambda}{\cos\theta} \frac{\partial \left(n_{eff,core}^{i} + n_{eff,clad}^{i} \right)}{\partial n_{s}} + \frac{n_{eff,core}^{i} + n_{eff,clad}^{i}}{\cos\theta} \frac{\partial \Lambda}{\partial n_{s}} \right) \Delta n_{s} \quad (13)$$

Due to $\partial\Lambda/\partial n_{s} = 0$, and noting that $\partial n_{eff}/\partial n_{s} = 0$ and $\partial n'_{eff}/\partial n_{s} = 0$ for the standard single mode telecommunication fibers such as CORNING SMF-28 fibers, we obtain:

$$\Delta\lambda_{B} = 0 \qquad\qquad (14)$$

$$\Delta\lambda_{clad}^{i} = \frac{\Lambda}{\cos\theta} \frac{\partial n_{eff,clad}^{i}}{\partial n_{s}} \Delta n_{s} \qquad\qquad (15)$$

This gives out the obvious result that only the cladding mode resonance wavelength shifts depend on the surrounding refractive index perturbations and that this dependence is scaled by the dispersion of the cladding mode effective index with the outside index.

3. Experimental Results

The experimental TFBG sensors are written in hydrogen-loaded CORNING SMF-28 fibers using 193 nm ArF excimer laser light or 248 nm KrF excimer laser light with a phase mask to generate the grating pattern. The setup for writing TFBGs is shown in Figure 2. In this configuration, the internal tilt angle θ of the grating planes can be calculated from the mask tilt angle α by applying Snell's law of refraction at the air-glass interface of the fiber. The resonance wavelengths of the Bragg and cladding modes are determined by the projection of the grating period of the mask used on the fiber axis and on the effective indices of each mode. Figure 3 shows the transmission spectrum of a TFBG written in a SMF-28 fiber with tilt angle $\theta=4°$ using 193 nm ArF excimer laser light. An important feature of the small tilt angle

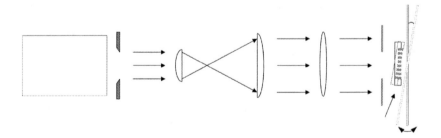

Figure 2. The top view of setup for TFBG written.

TFBG is the presence of a strong "ghost mode" resonance, immediately to the left of the Bragg resonance. This ghost mode is made up of several low order cladding modes and is known to be very sensitive to bending[3], but very little to outside refractive index[9]. The spectrum in the shorter wavelength region, which is about 20 nm away from the core resonance shows that in this region the cladding resonances become double and triple peaked. This is due to the fact that for higher order cladding modes, the fields are less confined in cladding layer and therefore the degeneracies of the LP mode approximations are split up into individual full vectorial modes of the fiber[18]. The inset of Figure 3 shows the transmission spectrum of the short wavelength region in detail.

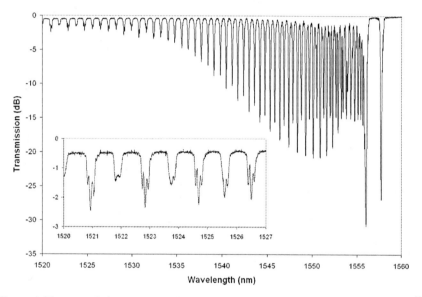

Figure 3. The transmission spectrum of 4° tilted FBG written in SMF-28 single mode fiber.[11]

3.1. DIFFERENTIAL STRAIN SENSITIVITY OF TFBG

Figure 4 shows the relative wavelength shifts of cladding resonances with respect to the Bragg resonance when longitudinal strain is applied to a TFBG in standard SMF-28 fiber with cladding diameter d=125 μm and θ=4° (a) written by 193 nm excimer laser; and (b) written by 248 nm excimer laser light. It can be seen for the strain perturbations that there are 3 differential wavelength shift regions. First, the ghost mode region appears to be sensitive to external strain perturbations with some low order cladding modes having negative relative wavelength shifts (i.e. cladding modes shift

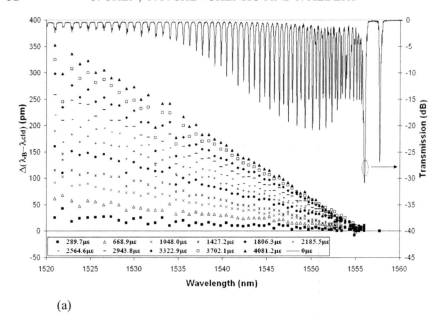

(a)

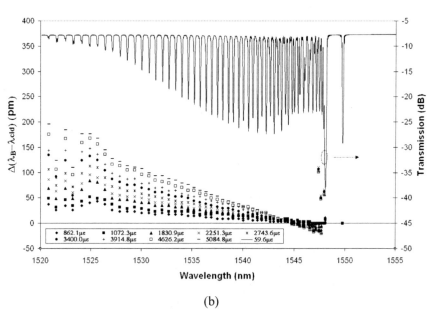

(b)

Figure 4. The transmission spectrum of 4° tilted FBG and differential wavelength shifts due to strain changes: (a) written by 193 nm ArF excimer laser light[11]; (b) written by 248 nm KrF excimer laser light.[12]

more than the core mode). For higher order modes, further than 20 nm from the Bragg resonance, the results become irregular because the cladding mode resonances have double- and triple-peaked resonances which are difficult to follow reliably over large shifting ranges. If the same individual peak of a multi-peak resonance could be followed unambiguously, then it is likely that the differential wavelength shift should continue to be linear in that region of the spectrum. Finally, between 5 nm and 20 nm from the Bragg resonance there is a wavelength region where the differential wavelength shift grows very linearly with mode order and with strain. This region of the spectrum will be the most useful for temperature independent strain sensors.

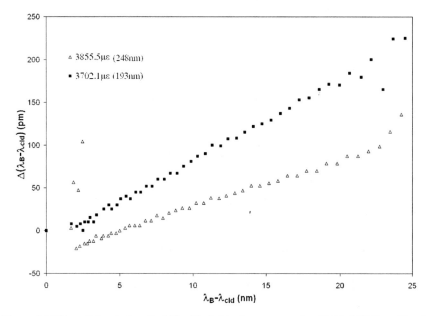

Figure 5. Differential wavelength shift of individual resonances for 4° tilted FBGs written by ArF and KrF excimer laser light.[11]

Figure 5 gives the comparison of the differential strain sensitivity of two 4° tilted FBG sensors written in SMF-28 single mode fiber by 193 nm ArF excimer and by 248 nm KrF excimer laser light respectively for a high strain value in each case (3702 με and 3855 με). Apart from irregular regions at both ends of the mode spectra in each case, the TFBG written at 193 nm shows a much larger differential strain. This result points to a yet unknown difference in the photosensitive mechanisms for the two types of irradiation. One hypothesis is that it has been shown previously that 193 nm irradiation of similar fibers to those used in this study causes significant compaction of the glass (more so than using 248 nm irradiation). This change in the physical structure of the glass could be sufficient to modify the photo-elastic coefficients p_{11} and p_{12} and hence the differential strain sensitivity of the TFBGs[19].

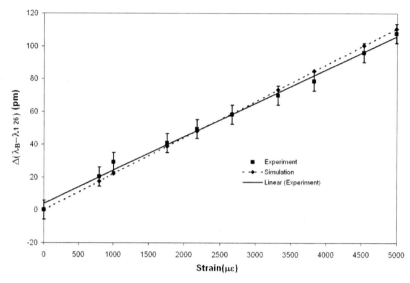

Figure 6. The relative wavelength shift of $LP_{1\,26}$ mode to $LP_{0\,1}$ mode for different strains with maximum temperature error.[10]

Figure 6 shows the experimental and analytical results for the differential wavelength shift of mode LP_{126} which has a resonance wavelength ~1530 nm relative to the LP_{01} resonance as a function of strain corresponding to Figure 4(b). The error bars on the experimental result reflect the maximum temperature error ~±6 pm for ~80°C temperature change that is observed in the wavelength shift region that shows regular behavior (see Section 3.2). The analytical result is obtained by equations (5) and (6), where the partial derivatives of all the effective indices with respect to strain have been replaced by the value for bulk silica, obtained from (9) with $p_{11} = 0.121$ and $p_{12} = 0.27$, and Poisson's ratio $v = 0.17$[20]. It can be seen that even with this approximation, the analytical and experimental results are in a good agreement (the small remaining difference reflecting the effect of the dopant materials and waveguide dispersion on the partial derivatives making up (5) and (6)).

3.2. DIFFERENTIAL TEMPERATURE SENSITIVITY OF TFBG

We now demonstrate experimentally that the temperature sensitivity can be factored out of the measurements by using the differential wavelength shift (with the Bragg resonance as reference). The temperature sensitivity of the differential wavelength shift of cladding modes relative to the core mode wavelength shift is shown in Figure 7 for the same TFBGs shown in Figure 4. The experimental results show that within the temperature range from –10°C to 70°C, the relative cladding mode shifts are less than ±12 pm or equivalent to an apparent ±1.2°C drift in over ~80°C of temperature variation (except for the short wavelength region) for both TFBGs. Therefore, within this uncertainty and this temperature range at least, a temperature insensitive sensor can be made by monitoring the relative core and cladding mode wavelength shifts in the transmission spectrum.

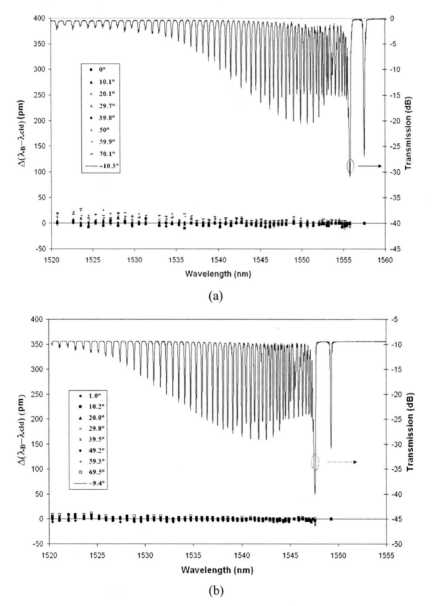

(a)

(b)

Figure 7. The transmission spectrum of 4° tilted FBG and differential wavelength shift due to the temperature changes: (a) written by 193 nm ArF excimer laser light[11]; (b) written by 248 nm KrF excimer laser light.[12]

Figure 8 shows how a single cladding resonance wavelength (near 1536 nm) shifts with respect to strain and temperature. From the results, we can see that the differential wavelength shift increases almost linearly as the strain increases, while the differential wavelength shift due to the temperature changes is within ±12pm in a range of 80°C. The strain sensitivity of the differential wavelength shift is 49 nm/unit strain (0.049 pm/με), while the temperature sensitivity of the same resonance is approximately 0.3 pm/°C with some degree of randomness. If we interrogate this resonance without independent knowledge of temperature, the error due to temperature fluctuations corresponds to an uncertainty of ±250 με.

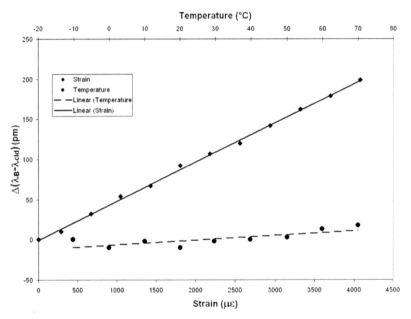

Figure 8. Comparison of the differential wavelength shifts of a single cladding mode resonance for a 4° tilted FBG at wavelength ~1536 nm due to strain and temperature perturbations.[11]

3.3. SURROUNDING REFRACTIVE INDEX SENSITIVITY

The experimental TFBG used was written by KrF excimer laser light with 6° tilted angle, its transmission spectrum measured in air is shown in Figure 9. The first set of results pertains to the measurement of sugar water solutions. The Figures 10 (a) and 10 (b) present the TFBG overlapped spectra for several values of n_D (refractive index measured at 589 nm with an Abbe refractometer) ranging from 1.377 to 1.43 in sugar water solutions. There is a notable shift of the high order cladding modes while lower order modes and the Bragg resonance are not affected. The resonance shown in Figure 10c is located about 15 nm away from the Bragg resonance. It is clear that this cladding mode resonance shifts by amounts that are easily measurable (with respect to the width of the resonance, ~100 pm FWHM). This single resonance can be used to measure media with n_D values between 1.0 and above 1.43 without ambiguity from neighboring resonances since the spacing between resonance is larger than 600 pm while the resonance shifts by 500 pm over this range of n_D. If we now plot the relative resonance position (with respect to λ_B) as a function of n_D shown in Figure 11, we see that there is a one-to-one relationship between SRI and resonance shift (another resonance located near 17 nm away from the Bragg resonance was used here). A seventh-order polynomial fit of the experimental results can be used as a calibration to correlate the measured wavelength shift with n_D. However, the sensitivity is quite poor for low values of SRI (20 pm/u.r.i. near $n_D = 1.38$) and increases dramatically as the SRI approaches the value at which this particular cladding mode becomes cut-off (3900 pm/u.r.i. near $n_D = 1.43$). This sensitivity increase near cut-off is typical of all sensing mechanisms that rely on the penetration of the evanescent field of the cladding modes into the outer medium[21]. However, this is where the TFBG configuration becomes advantageous since one can select different resonances of the same sensor, depending on the refractive index range of interest, to achieve the highest sensitivity. If the refractive index range of a particular situation is not known beforehand, then a simple pattern analysis of the transmission spectrum (compared to a stored transmission spectrum in air or in pure water) will reveal immediately which modes approach cut-off and where the measuring wavelength window should be located relative to the Bragg wavelength (for maximum sensitivity). By contrast, with an LPG-based

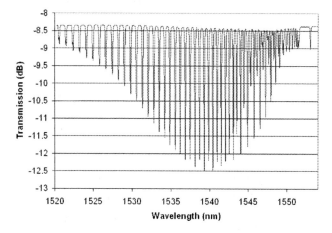

Figure 9. Typical experimental TFBG transmission spectrum (CORNING SMF28 fiber, $\theta = 6°$) measured in air.

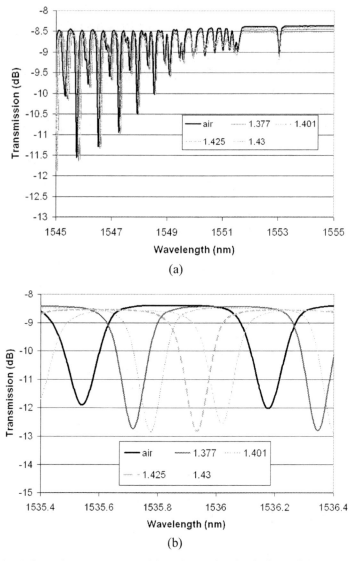

Figure 10. (a) Several measurements with various refractive indices of the outer medium, near the Bragg resonance. (b) Same spectra as (a) but zooming in on a particular resonance near 1535.5 nm.

sensor only a few (and generally only one or two) cladding modes are visible in a measurement window covering several hundred nm. In such cases a new sensor design (grating period) is required to optimize the sensitivity over different refractive index regions.

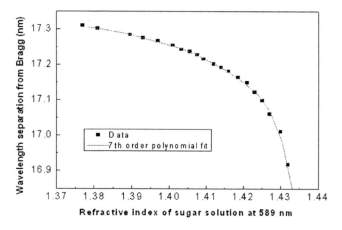

Figure 11. Experimental shift in the distance of a cladding mode resonance from the Bragg wavelength as a function of the refractive index of a sugar solution at 589 nm.

Further tests were carried out with calibrated liquids to demonstrate how quantitatively we can predict theoretically the refractive index sensitivity of several mode resonances and also to find out the temperature cross-sensitivity of our sensors. Figure 12 shows the comparison between predict-ted and measured differential wavelength shifts as a function of surrounding refractive index for various cladding mode orders. The cladding modes are identified here by the value of $(\lambda_B-\lambda_C)$ measured when the grating is in air (i.e. the original distance of the resonance from the Bragg wavelength irrespective of the mode identity and symmetry). The predicted shifts (solid lines) were obtained with a commercially available software package[22] using the refractive index profile and geometry of the standard fiber used. The fiber parameter used are: core radius = 4.15 μm, cladding radius = 62.5 μm, n(core) =1.450699, n(clad)=1.444024, both at 1566 nm and corrected for dispersion for other wavelengths, using the dispersion of pure silica as a guide and offsetting the dispersion curves to the refractive indices given above for λ = 1566 nm. When experimental error is taken into account all the measurements agree with values predicted from the model. For single measurements, we use an error estimate of +/–12 pm for the exact value of the wavelength distance between a resonance and the Bragg wavelength relative to its value in air, to reflect an absolute uncertainty of at least 3 pm on each wavelength measured. For some of the measurements shown in Figure 12, five independent measurements were made with the same liquid, and the average value of the wavelength distance was calculated. In such cases the error estimate is provided by the standard deviation of the

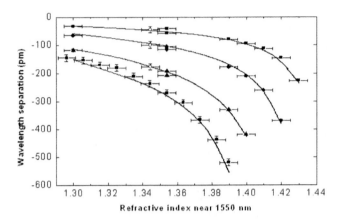

Figure 12. Change in wavelength separation from the Bragg wavelength for four different cladding modes as a function of the refractive index of calibrated liquids Filled symbols = experimental data at room temperature (23°C); Open symbols = Experimental data at 50°C; Lines = simulations.

measurements. The standard deviation for these measurements never exceeded 3 pm. The uncertainty in the value of the refractive index is given as +/–0.0002 (at 589 nm) by Cargille. For longer wavelengths the supplier provides Cauchy coefficients to calculate the dispersion of the refractive index but with an added uncertainty of the order of 0.005. Our results seem to indicate that the actual uncertainty of the refractive index of the liquids is much smaller.

In order to validate our claims of relative temperature independence, an additional measurement was carried out at a temperature of 50°C (open circles on Figure 12). When the thermo-optic coefficient of the liquid (-3.41×10^{-3}/°C) is taken into account, the refractive index values obtained from the grating also fall on the expected curve. This indicates that the sensor was able to measure the true refractive index of the liquids (within error), even though all its wavelength resonances were shifted by about 265 pm due to the increase in temperature (+27°C). The sensitivity of the SRI measurements (and simulations) reaches over 10000 pm/u.r.i. for each of the resonances shown in Figure 12, corresponding to better than 10^{-4} refractive index sensitivity (u.r.i./pm) between 1.38 and 1.43. The sensitivity calculated here for single resonance measurements (10^{-4}/pm) obviously needs to be adjusted by the wavelength measurement accuracy (about 10 pm for the relative shift) to yield an actual sensitivity closer to 10^{-3} in SRI in our experiments. On the other hand, modern dedicated fiber grating interrogation equipment claims a wavelength measurement accuracy better than 1 pm[23], making 10^{-4} sensitivity a definite possibility. With such an instrument, we have achieved a repeatability of the measurement of the wavelength distance (standard deviation of successive measurements of a fixed experimental situation) of 0.6 pm.

If there is a need for higher sensitivity at lower values of n_{ext}, there are two design modifications that can be used to do so. First, it is possible to generate even higher order mode resonances with the TFBG, so that accurate resonance measurements can be made further away from the Bragg resonance. This can be achieved with a larger tilt angle, as shown in Figure 13 where cladding mode resonances as far away as 100 nm from the Bragg wavelength are observed for tilt angles near 10 degrees. These resonances have effective indices near 1.25 and hence allow high sensitivity measurements in this range of values of the SRI. With the grating shown in Figure 13, we have achieved 11200 pm/u.r.i. for a SRI near 1.31 by using the resonance at 1500 nm (see Figure 14). The additional benefit of using resonances further away from the Bragg wavelength is that their spacing increases (to over 1200 pm in this case), thereby augmenting the range of unambiguous readings for the position of the resonance relative to its position in air (a similar effect occurs with reduced diameter cladding fibers[12] but require custom fabrication of the fibers or thinning methods lead to reduced mechanical strength). A second way to increase sensitivity in the lower index range is to use a thin layer of high index material on the cladding to increase the extent of the evanescent field of certain cladding modes into the outside medium. This may lead to an increased overlap and sensitivity to SRI changes, as was shown in the case of LPGs[24–25]. A major contrasting point between our TFBGs and devices based on LPGs, a large number of cladding modes is always available with a single grating device design, and that the exact wavelength of the Bragg resonance of the device

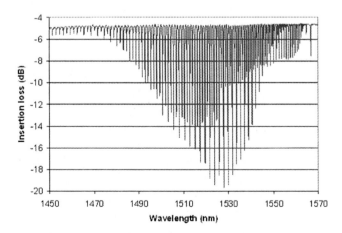

Figure 13. Experimental transmission spectrum of a TFBG with a tilt angle of 10 degrees, (λ_{Bragg} = 1566.810 nm).

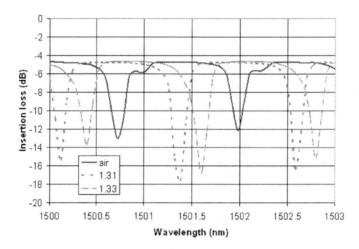

Figure 14. Shift of a resonance near 1500 nm for various external media (solid line: SRI=1.0, dotted line: SRI=1.3058, dashed line: SRI=1.3250, all values for λ=1500 nm)).

is quite irrelevant (as well as the exact strength of the grating!). These relaxed tolerances ensure that the mass production of such devices at very low cost is definitely possible.

Finally, the availability of several resonances within a single sensor measurement provides not only the possibility of multimodal sensing but also a way to increase the measurement accuracy of a given sensing mechanism by averaging the result obtained with the shifts of many cladding modes simultaneously. Since any systematic wavelength error is already eliminated by the fact that we are measuring relative wavelength shifts, the wavelength errors associated with each resonance should be statistically independent. Taking the results of Figure 12 as an example, we have four independent wavelength shift measurements for each value of SRI. We can calculate a predicted SRI from the simulation for each of these measurements, take the average and plot the results against the actual SRI as calculated from the supplier data and taking into account dispersion. The result is shown in Figure 15. In spite of the fact that the supplier only garantees +/–0.0052 accuracy on the refractive index of the liquids for wavelengths away from 589 nm, we obtain an excellent agreement between measured and expected values (as indicated by the slope of 0.999 and intercept of 5×10^{-4}) as well as a very good correlation (R^2=0.997). In fact, while the refractive index predictions from single measurements have errors ranging from –0.012 to +0.006, the errors obtained by averaging the predictions from 4 resonances range only from –0.0009 to 0.001. Even better results would be expected from a more sophisticated algorithm able to average the predictions of all the available resonances.

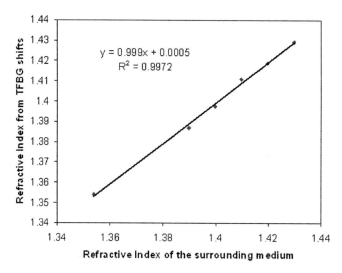

Figure 15. Measured refractive index from the average of the predictions based on the shifts of 4 resonances (–8 nm, –14 nm, –23 nm, and –28 nm) versus actual refractive index.

For comparisons, refractometers based on LPGs will have sensitivities of the order of 100–1000 nm/u.r.i. with cross sensitivity to temperature of the order of 1 nm/°C and above for the absolute measurement of the wavelengths of resonances with bandwiths ranging from 10-100 nm[24–26]. In the best cases, the sensitivity divided by the resonance bandwidth is 100 (u.r.i.)$^{-1}$ for LPGs, exactly equivalent to our results since it is just as easy to monitor pm shifts of 100 pm resonances as it is to monitor nm shifts of 100 nm resonances (easier in a way because the required spectral range of the interrogation systems is much smaller) and the same sensitivity enhancement techniques[24–26] can be applied to TFBGs as to LPGs. The only differences are that TFBGs have an inherent temperature monitoring channel (the Bragg resonance), and that they provide several measurements simultaneously to either achieve the highest sensitivity over several ranges of SRI, or to improve accuracy by averaging several measurements from a single sensor.

3.4. SURFACE PLASMON RESONANCES IN TFBG

In this case, we seek to determine whether the wavelength encoded cladding modes launched by the tilted fiber Bragg grating can be used to excite plasmon modes of a thin metal film (gold in this case) deposited on the cladding surface. The gratings were fabricated similarly as above.in CORNING SMF28 fiber. The transmission spectrum of the grating used for the experiments reported is shown in Figure 16. After fabrication, the gratings were heat-stabilized by subjecting them to a rapid annealing at ~300°C and the remaining hydrogen removed by 12 hours of heating at 120°C prior to gold deposition. In these preliminary experiments, the gold layers were deposited using a small-scale sputtering chamber (Polaron Instruments model E5100) with the fiber positioned a few cm from the gold target. For flat samples in the same geometry, a gold thickness of 20 nm requires 1 minute of deposition at a pressure of 0.1 Torr, a potential difference of 2.5 kV, and 18–20 mA of sputter current. In order to coat the fiber as uniformly as possible, two coating runs were made with the fiber holder rotated by 180 degrees between the coatings. Under these conditions, the film uniformity around the fiber circumference is unlikely to be very good and may lead to some polarization dependence of the light transmission. The film thickness on the fiber that we quote here is the value expected for the two sides of the fiber that are directly facing the sputtering target during the two coating runs. Profilometer traces were taken in samples where the gold film was scratched away to calibrate our process. While thicknesses ranging from 10 to 50 nm were tested, we concentrate in the following on results obtained with a 20 nm-thick nominal gold layer that actually measured 15 nm in the spots measured. We used a JDSU OMNI-2 Swept laser system to measure the transmission of our gratings: all the results shown were obtained by averaging the results for 4 orthogonal polarization states of the input light.

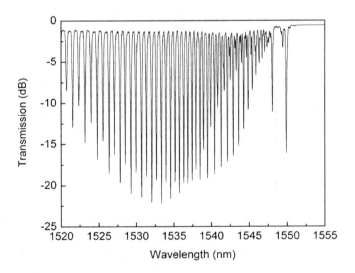

Figure 16. The transmission spectrum of a weakly tilted FBG.[14]

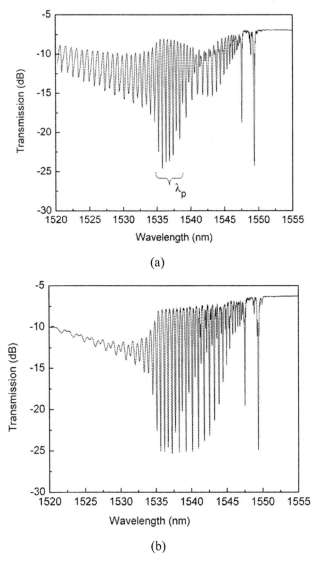

Figure 17. The transmission spectrum of the same grating as Figure 16 but immersed in a sucrose solution with n_D=1.4378: (a) with a 15 nm gold coating. The bracket indicates the peak position of the anomalous resonance[14]; (b) without gold coating.[14]

After the gold deposition, the fiber transmission spectrum is modified, but without measurable features of interest, indicating that the very thin gold layer has had an effect. When the gold-coated grating is immersed in liquids with various refractive indices (similar sugar solutions as above, still measured with an Abbe refractometer, for the refractive index of the solutions at 589 nm). It is clear from Figure 17a that anomalous resonances, completely different from any other TFBG spectra, appear for certain very specific sugar concentrations . The same grating in the same solution but without the gold film is shown on Figure 17b. The peak position of the anomalous resonance (λ_P on Figure 17a) is obtained by fitting the envelope of the cladding mode resonances. When the refractive index of the outer medium is changed gradually, the position of the anomalous resonance shifts is shown on Figure 18. The spectral width of the envelope of these anomalous resonances is about 5 nm.

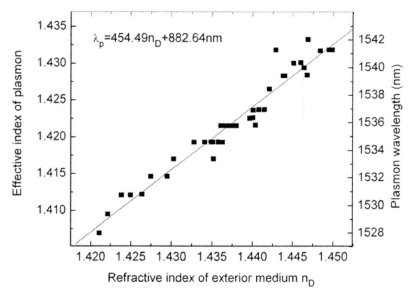

Figure 18. Dependence of the plasmon resonance peak wavelength (λ_P) and corresponding cladding mode effective index on the refractive index of the external medium at 589 nm (n_D).[14]

We now proceed to calculate some of the optical properties of these resonances in order to assess whether they correspond to those expected of plasmon excitations at the surface of the gold layer. We use equation (2) to find the effective indices of the cladding modes within a resonance and the refractive index of silica near 1550 nm (n=1.444) (we postulate that the effective index of the peak "anomalous" cladding mode is phase matched with a plasmon excitation) We can now calculate the angular spread of the equivalent angles of incidences (since the effective index is equal to the projection of the silica refractive index on the fiber axis). For the data of

Figure 17a, the angular spread is 3.5 degrees (around a mean incidence angle $\theta = 78°$). This angle of incidence agrees perfectly with the predicted value for gold-coated silica glass in sucrose solutions interrogated at wavelengths close to 1500 nm[27]. The angular spread of the resonance also corresponds well to typical values obtained for SPR measurements made with the Kretschmann configuration[28] (even taking into account that our deposition method likely results in non-uniform gold thickness around the fiber circumference, which would widen the resonances). Furthermore, the wavelength shift as a function of n_D is well approximated by a straight line with a slope of 454 pm/(10^{-3} change in n_D). Even considering the dispersion of our sugar solutions between 589 nm and the 1520–1560 nm region, this is again in excellent quantitative agreement with the expected behavior for contra-directional gratings in gold-coated silica fibers where shifts of the order of 100–500 pm/(10^{-3} change in next) were theoretically predicted[16]. These observations support our hypothesis that the resonance seen is indeed due to a SPR that is perturbing some of the cladding modes. In particular, the effective indices of the plasmons that we observe are smaller than the glass refractive index but larger than the effective indices of the outer medium. This corresponds to the situation described theoretically in Reference 5 where the plasmons are seen as perturbed cladding modes with a local electromagnetic field maximum at the outer metal boundary.

4. Summary

In this chapter, novel sensing mechanisms using TFBGs are analyzed theoretically and confirmed experimentally. The TFBGs can perform temperature-independent strain and refractive index sensing and, coupled with a nano-sized thickness gold coating, can serve as a platform for surface plasmon resonance sensing. Plasmon resonance sensors have been shown to be very promising for biochemical sensors due to their extraordinary characteristics and high sensitivity for surrounding medium.

Acknowledgement

Funding for this project was provided by the Natural Sciences and Engineering Research Council of Canada, the Canada Foundation for Innovation, and LxSix Photonics. J. Albert holds the Canada Research Chair in Advanced Photonics Components.

References

1. H. Sheng, M. Fu, T. Chen, W. Liu, and S. Bor, A lateral pressure sensor using a fiber Bragg grating, *IEEE Photon. Technol. Lett.*, 16, 1146–1148 (2004).
2. X. Shu, Y. Lin, D. Zhao, B. Gwandu, F. Floreani, L. Zhang, and I. Bennion, Dependence of temperature and strain coefficients on fiber grating type and its application to simultaneous temperature and strain measurement, *Opt. Lett.*, 27, 701–703 (2002).
3. S. Baek, Y. Jeong, and B. Lee, Characteristics of short-period blazed fiber Bragg gratings for use as macro-bending sensors, *Appl. Opt.*, 41, 631–636 (2002).
4. A. Martinez, Y. Lai, M. Dubov, L. Y. Khrushchev, and I. Bennion, Vector bending sensors based on fibre Bragg Gratings inscribed by infrared femtosecond laser, *Electron. Lett.*, 41, 472–474 (2005).
5. A. Iadicicco, A. Cusano, A. Cutolo, R. Berini, and M. Giordano, Thinned fiber Bragg gratings as high sensitivity refractive index sensor, *IEEE Photon. Technol. Lett.*, 16, 1149–1151 (2004).
6. A. Iadicicco, S. Campopiano, A. Cutolo, M. Giordano, and A. Cusano Microstructure fibre Bragg gratings: analysis and fabrication, *Electron. Lett.*, 41, 466–468 (2005).
7. S. C. Kang, S. Y. Kim, S. B. Lee, S. W. Kwon, S. S. Choi, and B. Lee, Temperature-independent strain sensor system using a tilted fiber Bragg grating demodulator, *IEEE Photon. Technol. Lett.*, 10, 1461–1463 (1998).
8. G. Laffont, and P. Ferdinand, Tilted short-period fibre-Bragg-grating-induced coupling to cladding modes for accurate refractometer, *Meas. Sci. and Technol.*, 12, 765–770 (2001).
9. C. Chen, L. Xiong, A. Jafari, and J. Albert, Differential sensitivity characteristics of tilted fiber Bragg grating sensors, *Proceedings of SPIE*, 6004, 6004–13 (2005).
10. C. Chen and J. Albert, Strain-optic coefficients of the individual cladding modes of a single mode fiber: theory and experiment, *Electron. Lett.* 48, 1027–1028 (2006).
11. C. Chen, L. Xiong, C. Caucheteur, P. Mégret and J. Albert, Differential strain sensitivity of higher order cladding modes in weakly tilted fiber Bragg gratings, *Proceedings of SPIE* 6379, 63790E1-7 (2006).
12. C. Chen, C. Caucheteur, P. Megret, and J. Albert, Sensitivity of Tilted Fiber Bragg Grating Sensors with Different Cladding Thicknesses, *18th Conference on Optical Fiber Sensors*, TuE31, (2006).
13. C. F. Chan, C. Chen, A. Jafari, A. Laronche, D. J. Thomson and J. Albert, Optical fiber refractometer using narrowband cladding mode resonance shifts, accepted for publication in *Appl. Opt.* (2006).
14. Y. Shevchenko and J. Albert, Plasmon Resonances in Gold-Coated Tilted Fiber Bragg Gratings, *Opt. Lett*, 32, 211–213 (2007).
15. Y.-J. He, Y.-L. Lo, J.-F. Huang, Optical-fiber surface-plasmon-resonance sensor employing long-period fiber gratings in multiplexing, *J. Opt. Soc. Am. B* 23, 801–811 (2006).
16. G. Nemova and R. Kashyap, Fiber-Bragg-grating-assisted surface plasmon-polariton sensor, *Opt. Lett.* 31, 2118–2120 (2006).
17. A. Othonos and K. Kalli, *Fiber Bragg Gratings*, (Artech House, Boston, 1999).
18. K. S. Lee and T. Erdogan, Fiber mode coupling in transmissive and reflective tilted fiber gratings, *Appl. Opt.*, 39, 1394–1404 (2000).
19. J. Albert, K. O. Hill, D. C. Johnson, F. Bilodeau, S. J. Mihailov, N. F. Borrelli, and J. Amin, Bragg gratings in defect-free germanium-doped optical fibers, *Opt. Lett.* 24, 1266–1268 (1999).

20. A. Bertholds and R. Dandliker, High-resolution photoelastic pressure sensor using low-birefringence fiber, *Appl. Opt.*, 25 340–343 (1986).
21. S. W. James and R. P. Tatam, Optical fiber long-period grating sensors: Characteristics and application, *Meas. Sci. and Technol.* 14, R49-R61 (2003).
22. OptiGrating 4.2, Optiwave Corp., http://www.optiwave.com.
23. High resolution swept laser interrogator Model Si720 from Micron Optics Inc. (Atlanta, GA).
24. A. Cusano, A. Iadicicco, P. Pilla, L. Contessa, S. Campopiano, and A. Cutolo, Cladding mode reorganization in high-refractive index-coated long-period gratings: effects on the refractive-index sensitivity, *Opt. Lett.* 30, 2536–2538 (2005).
25. J. Yang, L. Yang, C. Xu, C. Xu, W. Huang, and Y. Li, Long-period grating refractive index sensor with a modified cladding structure for large operational range and high sensitivity, *App. Opt.* 45, 6142–6147 (2006).
26. T. Allsop, F. Floreani, K. P. Jedrzejewski, P. V. S. Marques, R. Romero, D. J. Webb, and I. Bennion, Spectral Characteristics of tapered LPG device as a sensing element for refractive index and temperature, *IEEE/OSA J. Lightwave Technol.* 24, 870–878 (2006).
27. R. C. Jorgenson, S. S. Yee, A fiber-optic chemical sensor based on surface plasmon resonance, *Sens. Actuators B* 12, 213–220 (1993).
28. J. Homola, Present and future of surface plasmon resonance biosensors, *Anal. Bioanal. Chem.* 377, 528–539 (2003).

PHOTONIC LIQUID-CRYSTAL FIBERS: NEW SENSING OPPORTUNITIES

TOMASZ R. WOLIŃSKI
Faculty of Physics, Warsaw University of Technology
Koszykowa 75, 00-662, Warszawa, Poland

Abstract. The paper reviews and discusses the latest developments in the field of the photonic liquid-crystal fibers that have occurred for the last years in view of new challenges for both fiber optics sensing and liquid crystal photonics. In particular, we present the latest experimental results on electrically induced birefringence in photonic liquid crystal fibers and discuss possibilities and directions of future developments.

Keywords: Liquid crystals, photonic crystal fibers, induced birefringence

1. Introduction

In last few years there was a great interest in photonic crystal fibers (PCFs), which properties could be relatively easily changed after infiltrating with liquid crystals (LCs). Such a combination creates a new class of micro-structured fibers – photonic liquid-crystal fibers (PLCFs) in which thermal and electrical tuning possibilities along with unique spectral or polarization properties have been recently demonstrated. These microstructured PLCFs benefit from a merge of passive PCF host structures with "active" LC guest materials and could be responsible for diversity of new and uncommon propagation and polarization properties.

Over the last two decades anisotropic optical fibers have been extensively investigated[1] including also elliptical liquid-crystal core fibers that can reveal particular propagation and polarization properties.[2-5] Since 2003, there has appeared a new trend in liquid-crystal fibers research that relied on exploiting microstructured photonic crystal fibers in which air holes are filled with liquid crystalline materials[6]. In this way we obtained s a result a novel class of the microstructured fibers known as photonic liquid-crystal fibers[7-9] that could boost PCFs to a yet higher level of tunability.

W.J. Bock et al. (eds.), Optical Waveguide Sensing and Imaging, 51–72.

Photonic crystal fibers are created as an array of silica tubs and rods, which are then heated to around 2000°C and drawn down to the fiber. The core is usually made by a defect in the periodical structure of the PCF cross-section; it can be missing or additional rod as well as capillary. In this way either hollow-core fibers or solid-core fibers are manufactured. Guiding of the light in a PCF is governed by one of two principal mechanisms responsible for light trapping within the core. The first is a simple propagation effect based on the modified total internal reflection (TIR) phenomenon, which is well known and similar to the wave (index) guiding effect within a conventional fiber. The other is known as a photonic band gap (PBG) effect and it occurs if the averaged effective reflective index is lower in the core than in the cladding region. In this case the guiding mechanism relies on the coherent backscattering of light into the core. The PBG guiding is usually met in hollow-core photonic crystal fibers.

When a nematic liquid crystal is introduced into a glass capillary, its molecular orientation strongly depends on capillary dimensions, boundary conditions, and on physical fields influencing the anisotropic LC medium characterized by ordinary and extraordinary temperature-dependent refractive indices. Any external factor acting on the nematic core is in the position to change its effective refractive indices. The extraordinary index of refraction n_e depends on the angle α between propagation direction and the optical axis and is described by:

$$n_e(\alpha) = -\frac{n_o n_e}{\sqrt{n_0^2 \cos^2 \alpha + n_e^2 \sin^2 \alpha}} \tag{1}$$

where: n_o, n_e are ordinary and extraordinary refractive indices, respectively.

Highly birefringent (HB) optical fibers and their polarization properties have been extensively investigated for over the last two decades.[1,2] It is well known that in single-mode HB fibers linear birefringence lifts the degeneracy of the fundamental mode HE_{11} and two polarization modes have different propagation constants. Since HB (anisotropic) fibers have a pair of preferred orthogonal axes of symmetry, the two orthogonal quasilinear polarized field components HE_{11}^x and HE_{11}^y which propagate for all values of frequency, have electric fields polarized along one of these axes[1]. Hence, light polarized in a plane parallel to either axis will propagate without any change in its polarization, but with different velocities. However, injection of any other input polarization excites both field components HE_{11}^x and HE_{11}^y and as the two orthogonal mode components are characterized by different propagation constants β_x and β_y, they run into and out of phase at a rate determined by the birefringence of the HB fiber producing at the same time a periodic variation with a period L_B (beat length). Transmission properties of HB fibers in the fundamental mode are similar to those of anisotropic crystals in that the fiber has a pair of optical axes.

Unusual polarization properties including single-polarization index guiding by a multimode fiber have been observed in the elliptical-core fiber in which a hollow-core was filled with a low-birefringence liquid crystal.[3–5]

2. Photonic Liquid Crystal Fibers: Materials and Experimental Methods

First evidence of the band-gap tuning occurring with a PCF filled with high index oil was demonstrated in 2002 by Bise et al.[6] However, in the field of liquid crystals infiltration, first experimental demonstration of the performance of LC photonic band gap fibers was reported by Larsen et al. in 2003.[7] Electrical tuning in hollow-core PLCF was demonstrated in 2004[8] and in 2005 all-optical modulation in dye doped PLCF[9] was reported. Simultaneously, there has been interest both (theory and experiment) in possibility of birefringence tuning in different types of microstructured fibers[10] and to manipulate light in MOFs by microfluid motion.[11] HB tunable photonic band-gap fibers were numerically analyzed by Zhang et al.[12] and similarly PLCFs for single-polarization or high-birefringence guidance were theoretically modeled by Zografopoulos et al.[13]

We have recently experimentally demonstrated numerous propagation and polarization properties of solid-core PLCFs, mostly under the influence of temperature and electrical field in both visible and infrared wavelength regions.[14–19]

This chapter summarizes our experimental results on external field-induced effects (birefringence) in PLCFs. One of the approaches is to use as a host an isotropic PCF (manufactured in Lublin, Poland) that was filled in the length of few millimeters with different nematic LC mixtures of low (~0.05), medium (~0.15), and relatively high (~0.3) material birefringence synthesized at Military Univ. of Technology, Warsaw (Poland). Without the external electric filed, fiber birefringence is equal to zero due to the fact that liquid crystal molecules tend to orient parallel to the fiber axis. By switching electrical field perpendicular to the fiber long axis we observed a threshold molecular reorientation effect. This introduces changes in the effective refractive index of the holes infiltrated with LCs and induces birefringence in the PLCF cross section. As a result a change in transmission spectra and also an electrically-induced birefringence can be also observed It appeared that modal birefringence of the PLCF depends on both intensity of electrical field and material birefringence of the LC used to infiltrate the PCF.

3. Optical Properties in Liquid Crystals

LCs are organic materials generally consisting of molecules with long-range orientation order attributed to the van der Waals forces. Nematic liquid crystals have only long-range orientation order of the molecules long axes, but their centers of gravity are randomly distributed. Alignment of the nematic LC molecules, on the average, is characterized by one symmetry axis known as a director. Optical properties of the nematic liquid crystals differ for waves traveling normal or parallel to the director, and LCs are characterized by two refractive indices: ordinary and extraordinary that are strongly temperature dependent.

Unusual optical properties of liquid crystals are the background of their large-scale applications in liquid crystal displays, sensors, and optical data processing systems. The applications are mostly based on LC spatial light modulators (SLMs), devices that modify the amplitude, phase, or polarization of an optical wave front as a function of time and position across it. Since their discovery in 1888, LCs are generally classified into two classes: nematics and smectics but a few subgroups as chiral nematics and ferroelectric smectics are also distinguished due to their important optical properties and device applications. LCs are generally composed of elongated molecules that tend to align themselves under the effect of electric or magnetic fields. The birefringence of usual LCs is large, with a typical positive value ($\Delta n = n_e - n_o$) ~ 0.2 for a nematic LC in the visible region. Hence, the nematic LC phase behaves as a positive uniaxial medium ($n_e > n_o$). In the optoelectronic applications, the birefringence of the nematic LC sample is controlled by external applied electrical fields, which create phase shift, polarization switching, structure deformations, etc.

Optical properties of chiral nematics are very specific and are determined by helicoidal pitch (P), the arrangement of its axis, birefringence and the polarization of the incident light. These LCs are characterized by strong rotatory power, which manifests itself in rotation of linear polarization of light coming through a liquid crystalline film. LC films with high values of P are appropriate materials for e.g. pressure sensing since they possess strong rotatory power proportional to $1/P$ that can be easily modulated by deformation of film thickness. Sensitivity of such sensing devices depends on P, on birefringence $\Delta n = n_e - n_o$, and on elasticity of the LC layer. Chiral nematics are commonly used in displays such as twisted nematic (TN) cells and are thermally stable in a wide region of temperature. In a TN cell, a LC is introduced between two substrates treated to impose parallel molecular alignment but twisted with the angle 90°.

Light propagation in an anisotropic medium that twists about the direction of the propagating wave was originally discussed by Mauguin in 1911[1] but has received little attention until the re-discovery of liquid crystals in

'70s and the appearance of highly birefringent polarization maintaining (PM) optical fibers in '80s. In the so-called Mauguin limit that concerns the twisted anisotropic medium with large linear (intrinsic) birefringence dominating over the twisting effect, a linearly polarized incident wave injected along the birefringence axis remains linearly polarized and rotates at the same rate as the medium twists. For an oblique incidence, linearly polarized wave becomes elliptically polarized, the polarization ellipse rotates as the wave propagates through the medium, and the output polarization can be modified by twist rate and also by other perturbating effects as hydrostatic pressure, strain, temperature.[2]

As a host material for fabrication of the PLCFs presented in this chapter we used two different PCFs structures. The first one is a prototype, isotropic PCF manufactured at Maria Curie Skłodowska University (MCSU) in Lublin (Poland). The cross section of the PCF symbolically named as MCSU-1023 is shown in Fig. 1a. This structure contains a solid core surrounded with nine rings of air holes characterized by a diameter and hole spacing of 4.8 μm and 6.5 μm, respectively. The second host PCF is an anisotropic highly-birefringent and commercially available as *Blazephotonics* PM-1550-01 (Fig. 1b) PCF. In this fiber solid core is surrounded with two big holes of the 4.5 μm diameter and a set of smaller holes with diameter of 2.2 μm. PLCFs based on PM-1550-01 fiber could have only two big holes infiltrated with LCs, in contrast to the MCSU-1023 PCF in which each of the holes was LC infiltrated.

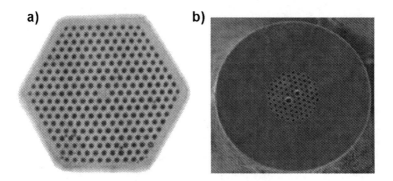

a)   b)

Figure 1. Photonic crystal fibers used as PLCF host matrix: a) MCSU 1023 PCF b) Blazephotonics PM-1550-01.

As an "active" element of the PLCFs we used LC of two different classes: with very low and medium material birefringence. Low birefringent LC prototype mixtures 1110 and 1550 were composed of alkyl 4-trans-(4-trans-alkylcyclohexyl) cyclohexyl-carbonates. They were synthesized according to the route shown in Fig. 2 presented in details elsewhere[20] and their optical properties were described by Schirmer et al.[21] Main difference between both 1110 and 1550 mixtures is their clearing temperature (transition temperature to the isotropic phase), which are 40.6°C and 78°C respectively.

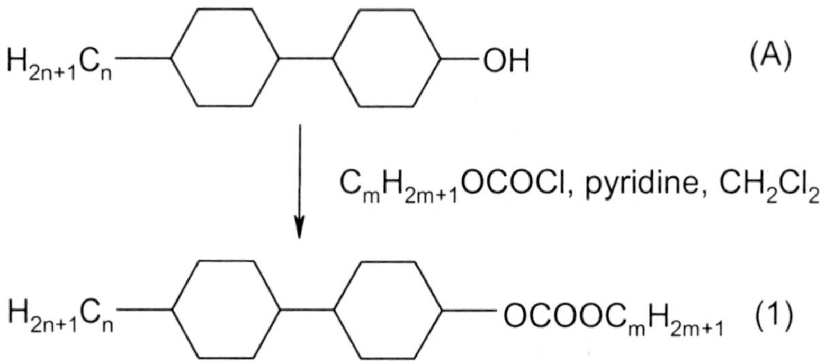

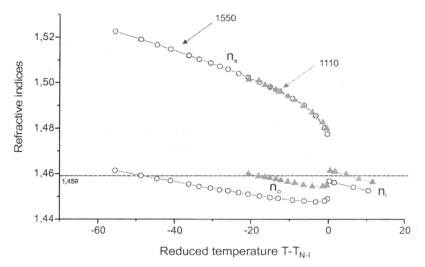

$$H_{2n+1}C_n \text{—} \bigcirc \text{—} \bigcirc \text{—OH} \qquad (A)$$

$$C_mH_{2m+1}OCOCl, \text{ pyridine, } CH_2Cl_2$$

$$H_{2n+1}C_n \text{—} \bigcirc \text{—} \bigcirc \text{—OCOOC}_mH_{2m+1} \quad (1)$$

Figure 2. Route of synthesis alkyl trans-4-(trans-4-alkylcyclohexyl) cyclohexylcarbonates.

Temperature dependence of refractive indices for the LC materials with very low birefringence (1110 and 1550) is shown in Fig. 3. These LC mixtures are especially interesting for silica glass fibers infiltrating, as their ordinary refractive indices in specific temperature ranges are lower than refractive index of the silica glass.

Figure 3. Refractive indices as a function of temperature for low birefringent LC mixtures: 1110 and 1550.

As LCs with medium birefringence we used commonly-known nematic LCs: PCB (4'-n-pentyl-4-cyanobiphenyl) and 6CHBT (4-trans-4-n-hexyl-cyclohexyl-isothiocyanatobenzene). Thermal characteristics of the refractive indices for these nematic LCs are shown in Figure 4. Each of the LC mixtures used for PCF infiltration was synthesized at Military University of Technology in Warsaw.

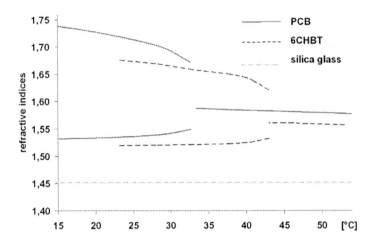

Figure 4. Refractive indices as a function of temperature for the medium-birefringence PCB and 6CHBT liquid crystals.

The 1550 liquid crystalline mixture was composed of alkyl 4-trans-(4-trans-alkylcyclohexyl)cyclohexylcarbonates (Table 1). They were synthesized according to the route shown in Figure 2 presented in details elsewhere[15]. Their optical properties were described in[14].

TABLE 1. Composition and mesomorphic properties of the 1550 LC mixture

R_1—⬡—⬡—R_2 wt.%	R_1	R_2	crystal-nematic phase transition temperature (°C)	nematic-isotropic phase transition temperature (°C)
23.93	C_3H_7	$OCOOC_2H_5$	59.4	71.4
18.24	C_5H_{11}	$OCOOCH_3$	70.5	95.0
34.55	C_5H_{11}	$OCOOC_2H_5$	54.8	85.0
23.28	C_3H_7	CN	57.0	79.7

One of the basic methods for manufacturing PLCFs by filling is to immerse an "empty" PCF in a container with a liquid crystal. Due to the capillary forces LC molecules are "sucked' into the holes and the filling speed depends mainly on LC viscosity, temperature and the current LC phase. The whole process can also be accelerated by using high-pressure air pumped into the hermetically sealed LC container. In our laboratory we usually filled ~10–30 mm long sections of a much longer (~50 cm) PCF. Then the manufactured PLCF was placed into a thermo-electric module allowing for temperature regulation in the 10–120°C range with ~0.1°C long-term stability and electric field regulation in the 0–1000 V range with frequency from 50 Hz to 2 kHz. As a light source we used high power halogen lamp, and output signal was analyzed by HR4000 fiber optics spectrometer (Ocean Optics). To investigate PLCFs polarization properties in the infrared range we used a tunable laser source (Tunics Plus CL, spectral range 1500÷1640 nm) and a modular system for polarization analysis PAT 9000B polarimeter (Tektronix).

4. Band-gap Propagation Tuning

A great majority of liquid crystals is characterized by both refractive indices higher than the refractive index of the silica glass used for the PCF fabrication process (Fig. 4). As a result in the PLCFs effective index of cladding is usually higher than the refractive index of the fiber core, so only propagation based on the PBG effect is possible. It means that propagation in the fiber core is possible only for wavelengths which correspond to photonic band gaps formed in cladding. Since optical properties of LCs can be relatively easy modified with external thermal, magnetic or electric fields, tuning of the photonic band gaps in the PLCFs is possible.

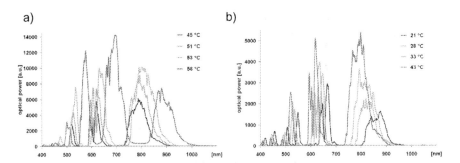

Figure 5. Thermal tuning in MCSU-1023 PCF filled with the 1702 LC mixture (Δn~0.3): a) red shift is observed if the ordinary refractive index is decreasing with temperature; b) blue shift could be observed if the ordinary refractive index is increasing with temperature.

Results of thermal bang-gaps tuning in the MCSU-1023 photonic crystal fiber filled with the medium-birefringence1702 LC mixture (~0.3) are shown in Fig. 5. Thermal dependences of its refractive indices have not been measured yet, however by analyzing the effect of thermally-induced PBGs tuning allows for qualitative description of the ordinary refractive index n_0 thermal characteristic. Since within the PLCF molecules are predominantly oriented along the fiber axis (flow-induced orientation) the band-gaps position is determined by the value of n_0, and if n_0 is decreasing a blue shift of PBGs is observed (Fig. 5a). Similarly, any increase of n_0 leads to a red shift of the PBGs (Fig. 5b).

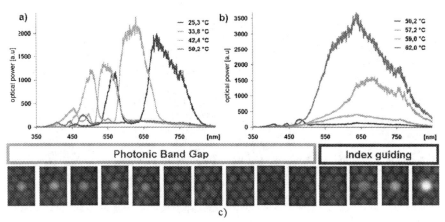

Figure 6. Switching between both guiding mechanisms in the PLCF filled with 1550 LC mixture: a) PBG propagation and band-gap tuning when ordinary refractive index of LC (n_{LCo}) is higher then refractive index of silica glass (n_{SiO2}); b) TIR guiding if $n_{LCo} < n_{SiO2}$; c) pictures of the end-face of the PLCF with PBG (left side) and mTIR (right side) propagation.

Very recently[19], an interesting effect was obtained in a PLCF filled with a LC characterized by extremely low birefringence and in which the ordinary refractive index – in a certain temperature range – is lower than the refractive index of the silica glass (Fig. 3). In such a situation the refractive index of PLCF core is higher than effective index of PLCF cladding, and the whole wavelength spectrum can propagate by the modified Total Internal Refraction mechanism. Figure 6 shows this thermally-induced and reversible switching between both guiding mechanisms: PBG and mTIR observed for the first time in the MCSU-1023 photonic crystal fiber filled with the 1550 low-birefringence LC mixture. Initially, a blue shift is observed (Fig. 6a) as n_0 is decreasing. No propagation is observed if n_0 matches exactly the refractive index of fiber silica core, but further temperature increase leads to propagation by the modified TIR mechanism (Fig. 6b).

It is quite obvious, that to achieve high-quality band gaps in the PLCF cladding the technological process of PCFs drawing should be well controlled to ensure that diameters of each hole will be equal. Even small fluctuations in the hole size have very destructive impact for PBG propagation. To demonstrate how these fluctuations affect PBGs we used a multi-core PCF, which cross section is shown in Fig. 7a. After infilling this fiber with 6CHBT we observed PBG propagation in every core; however in every core different spectra were guided (Fig. 7b). This indicates how important for manufacturing repeatable PLCFs with low attenuation is to use high-quality PCF structures.

a)   b)

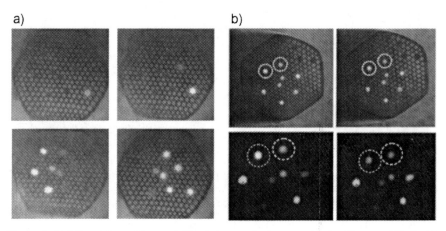

Figure 7. End-face of a multi-core PCF: a) the empty structure; b) the fiber was filled with the 6CBHT LC, different propagation spectra are visible in each of the cores as a result of hole diameters fluctuations.

5. Electrically Tunable Polarizing Fiber

Electrical tuning in PCF filled with liquid crystal has been recently reported in hollow-core fibers[9] and solid-core fibers[10,11]. In this section we briefly present experimental results similar to those reported elsewhere[9,10], in view of introducing a new method of single polarization propagation that will be referred to in next sections.

In the PCF filled with the 6CHBT nematic liquid crystal only selected wavelengths can be guided by the Photonic Band Gap mechanism. PBGs positions in the PLCF depend on the effective refractive index of the LC-infiltrated cladding holes and hence can be strongly sensitive to temperature changes, but also depend on molecules orientation. Examples of possible LC molecular alignments within the PLCF are shown in Fig. 8.

The simplest way for dynamical modification of LC molecular orientation is to use an external electric field. Transmission spectra of the MSCU-1023 fiber infiltrated with the 6CHBT nematic LC for different values of the electric field applied perpendicularly to the PLCF axis are shown in Fig. 9. Below a threshold voltage (~100 V) propagation is governed by the ordinary refractive index within the LC capillaries (Fig. 9a), however above this value molecular reorientation occurs and PBGs positions depend on the LC extraordinary index (Fig. 9b).

Electrically-induced molecular tilt introduces not only changes in the effective refractive index of the holes infiltrated with LCs, but also creates anisotropy in the PLCF cross section. In the off-voltage state, fiber birefringence is equal to zero due to the fact that liquid crystal molecules tend to orient parallel to the fiber axis (Fig. 8a). Any change in the azimuth of linearly polarized light launched into the PLCF causes a slight modulation of the optical power transmitted trough the fiber.

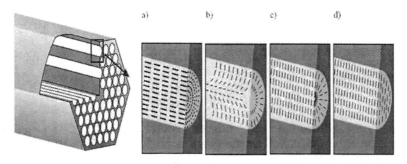

Figure 8. Examples of the liquid crystal molecules orientation within the holes of microstructured fibers: a) planar; b) radial; c) axial; d) transverse that introduces anisotropy in the PLCF.

Under the influence of the external electric field a threshold molecular reorientation occurs and at voltages higher than the threshold value, LC molecules tend to reorient perpendicularly to the fiber axis (Fig. 8d). As a result a change in transmission spectra and also in an electrically-induced birefringence can be observed. In this case by changing the azimuth of the linearly polarized light coupled into the PLCF we can obtain almost 100% power modulation (Fig. 10). It means that by using the electric field we are able to realize dynamic switching between an isotropic fiber and a single-polarization anisotropic fiber. For the orthogonal polarization attenuation was so high that no propagation was observed. When the electric field was switched off the transmission spectra returned to that observed at 0 V. Relaxation time was not measured, however according to[10,11] we can assume that it should in the range of 50 ms.

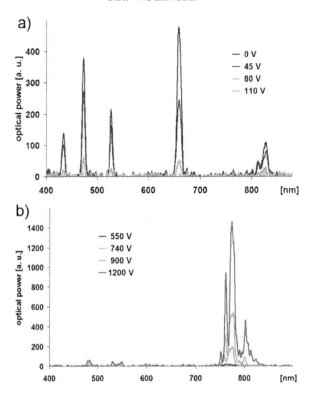

Figure 9. Electrically-induced tuning in the MSCU-1023 fiber with 6CHBT for different values of the electric field applied perpendicularly to the PLCF axis: below (a) and above (b) its threshold voltage.

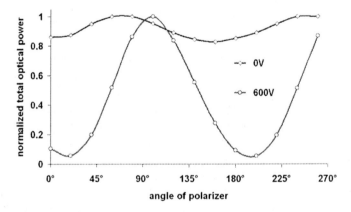

Figure 10. Modulation of the optical power transmitted in the PLCF through rotating the input polarization azimuth. At the off-voltage state (0 V) only small output power changes are observed. At higher voltages the PLCF becomes highly sensitive to the input polarization.

6. Birefringence: Inducing and Tuning

PBGs positions in a PLCF depends on the effective refractive index of the infiltrated with LCs holes. This value can be adjusted not only by temperature changes but also by molecules orientation modification. The simplest way to dynamical modification of LC orientation is applying an external electric field. Transmission spectra of the MSCU-1023 fiber with PCB for different values of the electric field (E-field) applied perpendicularly to the PLCF axis is shown in Fig. 11. Below a threshold voltage (~41 V), propagation is governed by the ordinary refractive index within the LC capillary, however above this value, the reorientation effect occurs and PBGs positions depend on the LC extraordinary index.

Electrically-induced molecules tilt not only introduces changes in the effective refractive index of the holes infiltrated with LCs, but also creates anisotropy in the PLCF cross section. In the off-voltage state, fiber birefringence is equal to zero due to the fact that liquid crystal molecules tend to orient parallel to the fiber axis. By changing the azimuth of the linearly polarized light launched into the PLCF cause only small modulation of optical power transmitted trough the fiber (this modulation results probably from the spectrometer grating polarization response).

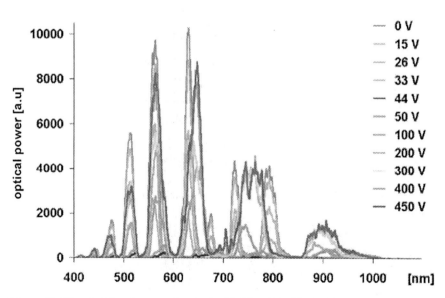

Figure 11. Electrical band-gap tuning in the PLCF (1023 filled with PCB). Above ~41 V propagation in the PLCF disappears, but further voltage increases induce propagation in new photonic band gaps.

By switching the E-field we observed a threshold molecular reorientation effect. As a result a change in transmission spectra and also an electrically-induced birefringence can be observed. In this case changing the azimuth of the linear polarization coupled into the PLCF leads to large (almost 98%) power modulation (Fig. 15).

Application of a highly birefringent PCF as a host for PLCF fabrication opens new possibilities for more advanced birefringence tuning. As an example, results of thermal birefringence tuning in the commercially-available PM-1550-01 HB fiber with two big holes filled with the 1110 LC is presented. Since the ordinary refractive index of the 1110 LC mixture is below the value of the core index (silica glass), so the whole spectrum is guided by the mTIR mechanism, however significant power modulation is observed (Fig. 12a). This effect occurs, as the spectrometer grating efficiency is different for both polarizations, and in effect the grating acts as a "partial analyzer". Therefore transmission maxima and minima can be observed, and distance between maxima depends on the PLCF birefringence. Transmission spectra for different temperatures are shown in Fig. 12b. The graphs have been limited to the 725–875 nm range in order to obtain better visualization of the maxima positions change. Since density of the maxima increases with temperature, birefringence of the PLCF also grows.

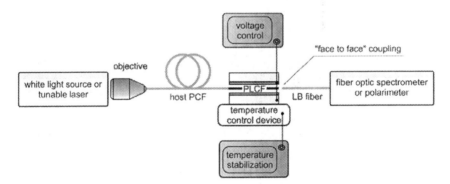

Figure 13. Experimental setup for investigation propagation and polarization properties of the PLCF.

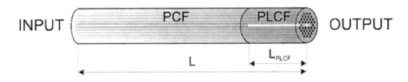

Figure 14. Configuration of the sample: the host Blazephotonics PM-1550-01 PCF (L~50 cm) in which only short section (L$_{PLCF}$ ~10÷30 mm) was infiltrated with a guest nematic LC.

Experimental details of the polarization measurements are presented in Figs. 13 and 14. The input light either from a broad-band source or from a tunable laser source was coupled into an empty section of the HB PCF. The terminal section of the PCF filled with LC i.e. the PLCF under investigation was placed between electrodes. The electrodes were plugged to a high-voltage source, which allowed for controlling both voltage from 0 V to 1000 V and frequency from 200 Hz to 2 kHz. The output optical signal coming out of the PLCF was analyzed by the Ocean Optics HR4000 fiber optic spectrometer. The measurement apparatus included also a tunable laser source (*Tunics Plus CL*) operating at third optical window (spectral range 1500÷1640 nm) and a modular PAT 9000B polarimeter (*Tektronix*) polarization analysis of the outgoing light from the HB PLCF. The PLCFs were thermally controlled by a temperature stabilization device. The Jones Matrix Eigenanalysis method was chosen for polarization mode dispersion measurements.

7. Photoalignment in PLCFs

Molecules orientation control in fiber holes is essential to reduce the PLCF losses and should guarantee repeatable results, which is especially impor-tant in context of potential mass production for commercial purposes. Infilling PLCFs by immersing PCF in containner with liquid crystal is relatively fast and easy method for PLCFs fabrication, but has one major disadvantage: poor control of randomly orientated molecules. It's seems obvious that more advanced methods of infilling should be developed. Some methods of molecules orientation control applied in LCDs technology (e.g. rubbing on polyimide layers) have limited application for holes with diameter of few microns.

Recently, we have been working on PLCFs with thin alignment film in which surface anisotropy is induced by photochemical reaction. Generally, photoinduced alignment could by achieved by exposing a photoalignment polymer with polarized or unpolarized ultraviolet (UV) light[25–29]. We have been using polyvinylcinnamate (PVCi) photopolymer, which molecules can be photocrosslinked through irradiation with linearly polarized UV light (Fig. 15).

Although preparation photoaligned PLCFs is much more complicated and time consuming, initial results are promising. In particularly significant difference has been observed in the MCSU-1023 fiber filled with 6CHBT and irradiated with UV light with different polarizations. In the PLCF irradiated with UV light with polarization direction perpendicular to the fiber axis high anisotropy in transmission spectra were observed, as a result of anisotropic molecules distribution in holes. After irradiation with UV light polarized parallel to the fiber axis good light propagation in holes were observed because LC molecules are well oriented along the fiber axis. Since experiments with photoalignment in PLCF are still in progress the more detailed results will be reported soon.

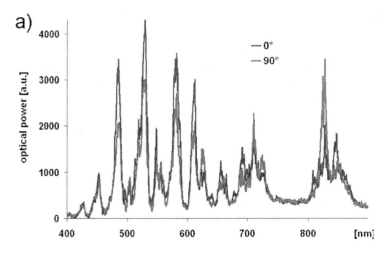

Figure 15. Photo-crosslinking between two PVCi molecules irradiated with linearly polarized UV light.

8.   PLCF Based on Highly Birefringent PCF

A highly-birefringent PCF (PM-1550-01, shown in Fig. 1b) used as a host fiber for the PLCF fabrication opens up a possibility for more advanced polarization properties tailoring. Due to anisotropic shape of its micro-structure, single-polarization propagation was observed just after all the fiber holes were filled with the 6CHBT nematic LC. It means that neither electrical field nor aligning layers are required for single polarization behavior of this PLCF. One of the possible explanations of this fact is that two orthogonally polarized modes have different mode profiles and their interaction with two large holes is also different. Since attenuation of liquid crystals is relatively high (especially when molecules orientation is not well defined), the mode that interacts stronger with the liquid crystal trapped in large holes is much more attenuated. The other possible explanation could be that positions of the photonic bandgaps must be polarization sensitive. However, for better understanding of this phenomenon extended numerical simulations have to be performed. In prospective simulations liquid crystal anisotropy has to be taken into account, as well as the fact that attenuation of the silica glass and LC-filled micro holes are significantly different. This seems to be a complex task and could be a subject of a future work.

Figure 16 shows transmission spectra for two orthogonal polarizations. For this PLCF a change in the input light polarization azimuth can lead to large (90–98%) polarization modulation that depends on the operating wavelength.

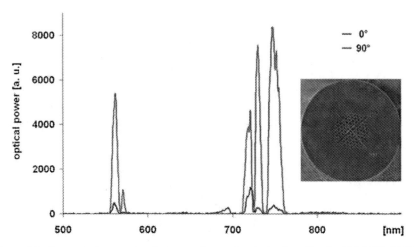

Figure 16. Transmission spectra for two orthogonal polarizations for the PM-1550-01 PCF filled with 6CHBT.

As an interesting example of thermal birefringence tuning in the PM-1550-01 PCF we present results obtained when only two large holes of this PCF were selectively filled with the low-birefringence 1110 LC mixture. For practical reasons we filled only 2 mm section of the 40 cm PM-1550-01 HB PCF. Since the ordinary refractive index of the 1110 mixture is below the value of the core index (Fig. 3), the whole spectrum is guided by the modified TIR mechanism, however significant power modulation can be recorded (Fig. 17a) in which transmission maxima and minima occur and the distance between maxima depends on the PLCF birefringence. Similar spectra can be observed if a highly birefringent fiber is placed between crossed polarizers [21]. However in our case we didn't use any analyzer at the output of the fiber and still we were able to observe the transmitted spectra modulation. It suggests that the HB PLCF itself can act as a "partial" polarizer eliminating need to use the output analyzer.

Let's denote by $\Delta\lambda$ a wavelength period that corresponds to a 2π phase shift between two polarization modes. The phase difference between two polarization modes induced by the PLCF of the length L is expressed as:[22–24]

$$\phi = \frac{4\pi L B}{\lambda} \tag{2}$$

where B_g is the group birefringence of the PLCF. For "conventional" highly birefringent fibers relationship between the group birefrin-gence and the wavelength (λ) is as follows:

$$B_g = -\frac{\lambda^2}{2L\Delta\lambda} \tag{3}$$

However in this case the measured sample was composed of two different highly birefringent fibers: the 38-cm long section of the empty PM-1550-01 fiber and the 20-mm long section of the PLCF. Consequently, the equation (3) could not be directly used for calculation of PLCF group birefringence and an interpretation of the results obtained is much more complicated since modal properties of both fiber sections (PCF and PLCF) are different.

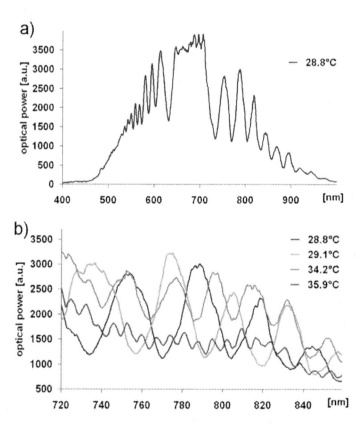

Figure 17. Thermal birefringence tuning in PLCF (PM-1550-01 PCF filled with the 1110 LC): a) transmission spectra at 25°C – whole spectrum guided by TIR mechanism, distance between maxima depends on value of PLCF birefringence; b) transmission in the 725–870 nm range at different temperatures, maxima density and the PLCF birefringence increases with temperature.

Moreover, even if birefringence of an empty PCF is well known it cannot be easily eliminated without extended numerical simulations. Equation (3) can be only used to estimate birefringence changes of the whole sample induced by temperature. Since the empty PM-1550-01 fiber is nearly thermally insensitive, so any birefringence changes can be attributed to the PLCF section. Transmission spectra for different temperatures are shown in Fig. 17b. The plots have been limited to the 725–875 nm range to visualize changes in the maxima positions. Since density of the maxima increases ($\Delta\lambda$ is decreases) with temperature, the group birefringence of the PLCF – according to the formulae (3) – should increase.

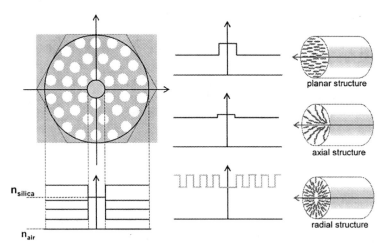

Figure 18. Possible "scenarios" of LC molecular ordering within the cladding holes of the solid-core PCF that can influence guiding mechanisms of the PLCF: planar and axial structures may cause mTIR effect whereas radial structure might be responsible for PBG phenomenon.

Interesting effects were envisaged when this PLCF sample was placed in the external electric field, since ordinary and extraordinary refractive indices of the 1110 LC mixture are respectively lower and higher than the refractive index of the silica glass. We expected that due to molecular reorientation one polarization could be guided by the modified TIR mechanism while the orthogonal by the PBG effect. Unfortunately, dielectric anisotropy of the 1110 LC is negative and molecules tend to align perpendicularly to the electric field direction. Since we have not observed any changes under the influence of the external electric field, so this indicates that the LC molecules were nearly perfectly oriented along the fiber axis. Further experiments involving the use of dual-frequency low-birefringence LCs as well as application of magnetic field are in progress.

9. Conclusions

Photonic liquid crystal fibers benefiting from a merge of passive photonic crystal fibers host structures with "active" liquid crystal guest materials create a new challenge for both fiber optics and liquid crystal photonics. They introduce new levels of tunability to photonic crystal fibers – the "hot topic" of modern fiber photonics and boost performance of these fibers due to diversity of new and uncommon propagation and polarization properties.

Apart of high sensitivity of the PLCFs to influences from external fields as temperature, electrical/magnetic/optical fields...the use of different molecular orientation "scenarios" within the micro holes (Fig. 18) can determine either index guiding (mTIR) or PBG propagation mechanism and reversible switching between them is possible.

In particular, we have demonstrated examples of thermally and electrically-induced birefringence changes in the PLCFs as well as the possibility of multi-core PLCFs. All these phenomena open up nearly unlimited possibilities of new applications in both optical signal processing and multi-parameter fiber-optic sensing.

Acknowledgements

The author is much indebted to many collaborators from Warsaw University of Technology, Military University of Technology, Maria Curie Skłodowska University in Lublin, Vrije Unversiteit Brussels – Professors: Roman Dąbrowski, Andrzej Domański, Mirosław Karpierz, Małgorzta Kujawińska – Doctors: Jan Wójcik, Edward Nowinowski-Kruszelnicki, Piotr Lesiak, Katarzyna Rutkowska, Tomasz Nasiłowski, Marek Sierakowski and Ph. D. students: Sławomir Ertman, Katarzyna Szaniawska, Aleksandra Czapla, Katarzyna Nowecka, and Urszula Laudyn for fruitful discussions and collaboration. A valuable technical assistance of Aleksandra Czapla, during the preparation of the manuscript is gratefully acknowledged. This work was partially supported by the Warsaw University of Technology and by the Polish Ministry for Science and Higher Education.

References

1. T.R. Woliński, " Polarimetric optical fibers and sensors", *Progress in Optics* XL, 1–75, edited by E. Wolf, North-Holland, Amsterdam (2000).
2. T.R. Woliński, in *Encyclopedia of Optical Engineering,* R.G. Diggers, ed., (M. Dekker, New York), 2150–2175 (2003).

3. T.R. Woliński, A. Szymanska, T. Nasilowski, E. Nowinowski, R. Dabrowski, "Polarization Properties of Liquid Crystal-Core Optical Fiber Waveguides", *Mol. Cryst. Liq. Cryst.*, Vol. 352, 361–370 (2000).

4. T. R. Woliński, A. Szymanska, "Polarimetric optical fibers with elliptical liquid-crystal core", *Measurement Science and Technology* 12, 948–951 (2001).

5. T.R. Woliński, P. Lesiak, R. Dabrowski, J. Kedzierski, E. Nowinowski-Kruszelnica, "Polarization mode dispersion in an elliptical liquid crystal core fiber", *Mol. Cryst. Liq. Cryst.* 421, 175–186 (2004).

6. T.T. Larsen, A. Bjarklev, D.S. Hermann, J. Broeng, "Optical devices based on liquid crystal photonic bandgap fibres", *Opt. Express* 11, 2589–2596 (2003).

7. R.T. Bise, R.S. Windeler, K.S. Kranz, C. Kerbage, B.J. Eggleton, and D.J. Trevor, "Tunable photonic band-gap fiber", *Optical Fiber Communication Conference Technical Digest*, pp. 466–468 (2002).

8. Y. Du, Q. Lu and S.T. Wu., "Electrically tunable liquid-crystal photonic crystal fiber", *Appl. Phys. Lett.* 85, 2181–2183 (2004).

9. T.T. Alkeskjold, J. Laegsgaard, A. Bjarklev, D.S. Hermann, J. Broeng, J. Li , S.T. Wu, "All-optical modulation in dye-doped nematic liquid crystal photonic bandgap fibers", *Opt. Express* 12, 5857–5871 (2004).

10. C. Kerbage, and B. Eggleton, "Numerical analysis and experimental design of tunable birefringence in microstructured optical fiber", *Opt. Express* 10, 246–255 (2002).

11. C. Kerbage, B.J. Eggleton, "Manipulating light by microfluidic motion in microstructured optical fibers", *Optical Fiber Technology*, Vol. 10, Issue 2, 133–149 (2004).

12. C. Zhang, G. Kai, Z. Wang, Y. Liu, T. Sun, S. Yuan, and X. Dong, "Tunable highly birefringent photonic bandgap fibers", *Opt. Lett.* 30, 2703–2705 (2005).

13. D.C. Zografopoulos, E.E. Kriezis, and T.D. Tsiboukis, "Photonic crystal-liquid crystal fibers for single-polarization or high-birefringence guidance," *Opt. Express* 14, 914–925 (2006).

14. T.R. Woliński, K. Szaniawska, K. Bondarczuk, P. Lesiak, A.W. Domanski, R. Dabrowski, E. Nowinowski-Kruszelnicki, J. Wojcik, "Propagation properties of photonic crystals fibers filled with nematic liquid crystals", *Opto-Electronics Review* 13(2), 59–64 (2005).

15. K. Szaniawska, T.R. Woliński, S. Ertman, P. Lesiak, A.W. Domanski, R. Dabrowski, E. Nowinowski-Kruszelnicki, J. Wojcik, "Temperature tuning in photonic liquid crystal fibers", *Proc. SPIE*, Vol. 5947, 45–50 (2005).

16. S. Ertman, T.R. Woliński, K. Szaniawska, P. Lesiak, A.W. Domanski, R. Dabrowski, E. Nowinowski-Kruszelnicki, J. Wojcik, "Influence of electrical field on light propagation in microstructured liquid crystal fibers", *Proc. SPIE*, Vol. 5950, 326–332 (2005).

17. T.R. Woliński, K. Szaniawska, S. Ertman, P. Lesiak, A.W. Domanski , R. Dabrowski, E. Nowinowski-Kruszelnicki, J. Wojcik, "Spectral and polarization properties of microstructured liquid crystal fibers", *Proc. SPIE*, Vol. 5936, 169–176 (2005).

18. T.R. Woliński, P. Lesiak, A.W. Domanski, K. Szaniawska, S. Ertman, R. Dabrowski, J. Wojcik, "Polartization optics of microstructured liquid crystal fibers", *Mol. Cryst. Liq. Cryst.* 454, 333–350 (2006).

19. T.R. Woliński, K. Szaniawska, S. Ertman, P. Lesiak, A.W. Domanski, R. Dabrowski, E. Nowinowski-Kruszelnicki, J. Wojcik, "Influence of temperature and electrical fields on propagation properties of photonic liquid crystal fibers", *Measurement Science and Technology* 17, 985–991 (2006).

20. W.J. Bock, A.W. Domański, T.R. Woliński, *Applied Optics.* 29, 3484 (1990).

21. J. Schirmer, P. Kohns, T. Schmidt-Kaler, A. Muravski, S. Yakovenko, V. Bezborodov, R. Dabrowski, P. Adomenas, "Birefringence and refractive indices dispersion of different liquid crystalline structures", *Mol. Cryst. Liq. Cryst.*, 307, 1–26 (1997).

22. M.W. Haakestad, T.T. Alkeskjold, M.D. Nielsen, L. Scolari, J. Riishede, H.E. Engan, A. Bjarklev, "Electrically tunable photonic bangap guidance in a liquid-crystal-filled photonic cystal fiber", *IEEE Phot. Techn. Lett.* 17, 819–821 (2005).

23. L. Scolari, T.T. Alkeskjold, J. Riishede, A. Bjarklev, D.S. Hermann, A. Anawati, M.D. Nielsen, P. Bassi, "Continuously tunable devices based on electrical control of dual-frequency liquid crystal filled photonic bandgap fibers", *Opt. Express* 13, 7483–7496 (2005).

24. J.A. Reyes-Cervantes, J.A. Reyes-Avandano, P. Helevi, "Tuning of optical response of photonic bandgap structures", *Proc. SPIE*, 5511, 50–60 (2004).

25. Dwight W. Berreman, "Solid Surface Shape and the Alignment of an Adjacent Nematic Liquid Crystal", *Phys. Rev. Lett.* 28, 1683–1686 (1972).

26. M. O'Neill and S.M. Kelly, "Photoinduced surface alignment for liquid crystal displays", J. Phys. D: *Appl. Phys.* 33, R67–R84 (2000).

27. V.V. Presnyakov, Z.J. Liu, V.G. Chigrinov, "Infiltration of photonic crystal fiber with liquid crystals", *Proc. SPIE* 6017, p. 60170J-1-7 (2005).

28. M. Schadt, K. Schmitt, V. Kozinkov, V.G. Chigrinov, "Surface-Induced Parallel Alignment of Liquid Crystals by Linearly Polymerized Photopolymers", *Japan. J. Appl. Phys.* 31, 2155–2164 (1992).

29. G.P. Bryan-Brown, I.C. Sage, "Photoinduced Ordering and Alignment Properties of Polyvinylcinnamates", *Liq. Cryst.*, 20, 825–829 (1996).

AN INTRODUCTION TO RELIABILITY OF OPTICAL

COMPONENTS AND FIBER OPTIC SENSORS

FRANCIS BERGHMANS[*]
SCK·CEN, Boeretang 200, 2400 Mol, Belgium and
Vrije Universiteit Brussel, Pleinlaan 2, 1050 Brussels,
Belgium

SOPHIE EVE
ESTCM du CRISMAT, 6 Boulevard Maréchal Juin,
14050 Caen Cedex 4, France

MARCEL HELD
EMPA, Materials Science and Technology, Ueberlandstrasse
129, 8600 Duebendorf, Switzerland

Abstract. We shortly review general reliability engineering concepts and methods and attempt to discuss in how far these can be applied to optical components used for optical fiber sensors.

Keywords: Reliability; optical components; optical fiber; optical fiber sensor; Bragg grating

1. Introduction

Optical fiber sensors have been deployed for many years in a wide variety of applications. Each of these applications comes with its own specific demands to which optical engineers have always found resourceful answers. As a result a multitude of optical sensing methods have been developed and many different sensors have been designed, manufactured and applied, some of them very successfully. These sensors are built using diverse kinds of optical fibers and waveguides. The latter are sometimes modified in a dedicated way, for example by the inscription of Bragg gratings or by the

---

[*]Francis Berghmans, SCK·CEN, Advanced Reactor Instrumentation, Boeretang 200, B-2400 Mol, Belgium, fberghma@sckcen.be

W.J. Bock et al. (eds.), Optical Waveguide Sensing and Imaging, 73–100.
© 2008 *Springer.*

removal of parts of the fiber coating and sometimes even of the cladding. In the case of extrinsic fiber sensors, the fiber needs to be interfaced and packaged with an external sensing body, for example a Fabry-Perot cavity. This calls for the use of reflective end face coatings, epoxies, connectors, lenses, capillaries, chemical reagents, etc. Generally speaking, a fiber or waveguide sensor has to be considered as a complex system consisting of individual parts and components between an optical source and an optical detector. However, in the case of a distributed fiber sensor, it can be as "uncomplicated" as a single fiber between source and detector.

The issue of reliability of optical fiber and waveguide sensors becomes increasingly important as they are more and more frequently used in applications where a failure of the (often inaccessible) sensor might have dramatic consequences on cost and/or safety. An example of such an application is structural health monitoring. Assessing the reliability of a fiber sensor remains a complicated issue as this reliability can be defined on different levels:

- the reliability of the modulation that might require, for example, an optical source with very stable spectral and intensity characteristics;
- the reliability of the interaction between the measured physical quantity and the sensor probe, for example the strain transfer between the structure of interest and the strain sensing element;
- the reliability of the demodulation and signal processing, for example the accuracy and stability of wavelength measurements on fiber Bragg gratings;
- the reliability of the calibration or recalibration;
- the reliability on the component level.

This chapter intends to deal with the latter, i.e. component reliability. It is inspired from work carried out within a number of European networking initiatives, i.e. the former COST[*] 246 and COST 270 actions on "Reliability of Optical Fibers and Components" and "Reliability of Optical Components and Devices in Communications Systems and Networks", respectively and the European Union Framework Programme 6 Network of Excellence on Micro-Optics NEMO.

We first start by introducing general concepts and definitions of reliability engineering in section 2 in a way mainly inspired from Birolini[1] and from the Final Report of COST 246[2]. We then focus on what is known about the reliability of a number of selected optical components that are

---

[*]COST is the European Cooperation in the field of Scientific and Technological research and is supported by the European RTD Framework Programs.

typically part of a fiber sensor system, such as a the optical fiber itself and Bragg gratings in section 3. The choice for these examples is not innocent in the sense that they are probably the most extensively studied "passive" fiber optic sensor components. We chose not to deal with active devices such as light emitting diodes and laser diodes. The latter are often part of the interrogation or read-out instruments and can therefore be considered to be more easily replaceable and – in a first order approximation – to be of less importance for the overall reliability of a sensor system. Nevertheless one has to bear in mind that spectral and optical power stability might be very important requirements in a sensor network. Details on active optical device reliability can be found in the handbooks by Fukuda[3] and Ueda[4]. Section 4 then closes this chapter with considerations on the growing need for fiber optic sensing standardization and with some information on current activities in that domain.

As a final introductory remark, the authors would like to emphasize that this chapter has no intention whatsoever to give a full and detailed account of all the aspects related to optical component or optical fiber sensor reliability. Readers are therefore encouraged to consult the list of references selected with care for more extensive information on the different topics covered.

2. Principles of Reliability Engineering

Generally speaking a fiber optic sensor end user considers the "reliability" of the sensor as the "trustworthiness" with which a measurement is made and transparently interfaced to some kind of monitoring computer system. In the particular case of structural health monitoring for which reduced life cycle costs are of primary importance, the sensor should not just record a physical parameter in a dependable way. It should also be an element of an asset management technique that offers the potential of improved security, lower failure and environmental risk, extended service life, reduced maintenance costs and reduced downtime.[5] In this respect the reliability of the sensor is directly related to the operational effectiveness of the system. More generally reliability plays an important role in the concept of system or cost effectiveness (Figure 1).[1] System/cost effectiveness is a measure for the ability of an item to meet service requirements of defined quantitative characteristics with the best possible ratio of usefulness to life cycle cost and therefore is a prerequisite for profitable and sustainable technical systems in the market.

For a complex system, higher reliability generally leads to higher development costs and lower operating costs, so that the optimum life cycle cost is in between extremely low and high reliability figures. The objective of reliability engineering is to develop methods and tools to evaluate and demonstrate Reliability, Availability, Maintainability, and Safety (RAMS) of components, equipment, and systems, as well as to support developers and manufacturers to integrate these characteristics into their products. RAMS is of growing importance because of the increased complexity of equipment and systems and the high cost incurred by loss of operation. Equipment and systems have to be more than only free of defects when they are put into operation. The expectation is that they perform the required function without failure for a stated time and have a fail-safe behavior in case of critical failures. Whether an item will operate without failures can only be expressed by means of a probability. This probability is a measure for the item's reliability.

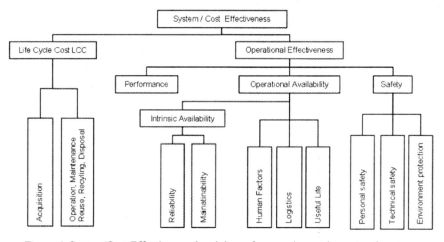

Figure 1. System/Cost Effectiveness breakdown for complex equipment and systems.

2.1. RELIABILITY

Reliability can be qualitatively defined as the ability of an item to remain functional. In quantitative terms, the reliability of an item (component, assembly or system) is the probability that this item will perform a certain function within a set of pre-defined specifications for a given period in time. It is generally referred to as *R*. A numerical statement of reliability (e.g. *R* = 0.9) must be accompanied by the definition of the required function, the operating conditions and the mission duration.

The required function specifies the item's task. For example, for given inputs, the item outputs have to be constrained within specified tolerance bands (performance parameters should be given with tolerances and not merely as fixed values). For example in the case of an optical component one might require that the emitted or transmitted optical power may not decrease over time to levels lower than a specified value, e.g. 50% of the initial optical power. The definition of the required function is the starting point for any reliability analysis because it also defines failures. The required function and/or operating conditions can be time dependent. In these cases, a mission profile has to be defined and all reliability statements will be referred to it. A representative mission profile and the corresponding reliability targets should be given in the item's specifications.

Operating conditions have an important influence on reliability and must therefore be specified with care. Experience shows for example that the failure rate of semiconductor devices will double for an operating temperature increase of 10–20°C.

In reliability theory, τ represents the failure-free operating time, thus τ is a nonnegative random variable, i.e. $\tau \geq 0$ and its distribution function $F(t) = 0$ for $t < 0$, or a positive random variable, i.e. $\tau > 0$ and $F(0) = 0$. $R(t)$ represents a survival function and expresses the probability Pr that an item will operate without failure in the interval $(0,t)$, usually with the assumption $R(0) = 1$, i.e. the item is operating when "switched on" at time $t = 0$. One can write the reliability function $R(t)$ as

$$R(t) = Pr\{\tau > t\} = 1 - F(t) \tag{1}$$

A more stringent assumption in investigating failure-free operating times is that at $t = 0$ the item is free of defects and systematic failures. Nevertheless, τ can be very short, for instance because of a transient event at turn-on.

A distinction has to be made between predicted and estimated or assessed reliability. The first one is calculated on the basis of the item's reliability structure and the predicted failure rate of its components while the second is obtained from a statistical evaluation of reliability tests or from field data with known environmental and operating conditions.

2.2. FAILURE

A failure occurs when an item stops performing its required function. Failures should be classified according to a mode, cause, effect and mechanism. The failure mode is the symptom (local effect) by which a failure is observed. Examples include opens, shorts or drift for electronic components; optical power loss or wavelength shift for optical components; brittle rupture, creep, cracking or fatigue for mechanical components. The cause of a failure can be intrinsic (early failure, failure with constant failure rate, and wear out failure) or extrinsic. Extrinsic causes often lead to systematic failures (due to errors, misuse or mishandling in design, production or operation) which are deterministic and should be considered like defects.

Defects are present – but often not manifest or discovered – at $t = 0$. The effect or consequence of a failure can be different if considered at the level of the item itself or at a higher level. A usual classification is: non-relevant, partial, complete, and critical failure. Since a failure can also cause further failures in an item, a distinction between primary failure and secondary failure is important. Finally, the failure mechanism is defined as the physical, chemical or other process resulting in a failure.

2.3. FAILURE RATE AND MEAN-TIME-TO-FAILURE

A most important reliability metric is the failure rate. The failure rate $\lambda(t)$ of an item showing a continuous failure-free operating time τ is defined as

$$\lambda(t) = \lim_{\delta t \to 0} \frac{1}{\delta t} Pr\{t < \tau \le t + \delta t | \tau > t\} \qquad (2)$$

or alternatively

$$\lambda(t) = -\frac{1}{R(t)} \frac{dR(t)}{dt} = -\frac{1}{N(t)} \frac{dN(t)}{dt} \qquad (3)$$

where $N(t)$ is the population of items that did not fail before a time t. The failure rate $\lambda(t)$ thus fully determines the reliability function $R(t)$. The failure rate of a large population of statistically identical and independent items typically follows a so-called "bath tub" curve as depicted in Figure 2.

When put into service one typically encounters early failures also known as the region of "infant mortality". These failures are due to weaknesses in the materials, components and/or production processes. In that region the failure rate decreases rapidly with time to reach a longer region with (nearly)

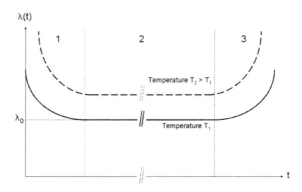

Figure 2. The typical evolution of the failure rate λ as a function of time. 1 – early failures, 2 – constant failure rate, 3 – wear out. The dashed curve represents what typically happens with the failure rate at a higher temperature.

constant failure rate. In case of screening and removal of the bad items before putting them into service, the failure rate will also be constant at early times. After a certain time the number of failures then starts to increase again due to wear out, aging and/or fatigue.

For many electronic and optical components one can assume the failure rate to be nearly constant and thus time independent, $\lambda(t) \doteq \lambda_0$. In that case the Reliability Function $R(t)$ is an exponential given by

$$R(t) = e^{-\lambda_0 t} \qquad (4)$$

The failure-free operating time is then exponentially distributed. In that case the number of failures is Poisson distributed with parameter $m = \lambda_0 t$. This is the only case for which one can estimate the failure rate as

$$\hat{\lambda}_0 = \frac{k}{T} \qquad (5)$$

where T is a given cumulative operating time and k the number of failures in that period. This stems from the fact that the exponential distribution function is memoryless, i.e. the property depends on the state at time t but not upon the history up to that time. In the general case, the mean or expected value of the failure free operating time $\tau \geq 0$ is given by the Mean Time To Failure *MTTF*

$$MTTF = E[\tau] = \int_0^\infty R(t)dt \qquad (6)$$

In the special case of a constant failure rate $\lambda(t) = \lambda_0$ this mean value $E[\tau]$ equals $1/\lambda_0$, which is usually designated as the Mean Time Between Failures *MTBF*.

$$MTBF = \frac{1}{\lambda_0} \qquad (7)$$

The failure rate is conventionally expressed in FIT (Failures In Time) where 1 FIT unit corresponds to 1 failure per 10^9 hours of operation.

2.4. FAILURE RATE ASSESSMENT

From the above it is obvious that the assessment of failure rates is a key issue in reliability engineering. One has to be aware that this is a difficult task because of the multitude and variety of electronic and optical components, the manifold of possible failure mechanisms and the pace of technological progress. In this section we briefly describe three generic methods for assessing failure rates: 1) the collection and the analysis of failure data from reliability tests, 2) physics of failure analysis and extrapolation of failure rates and 3) empirical prediction models based on the analysis of reliability field and test data.

Modern electronic or optical components failure rates are in the range of 10^{-10} to 10^{-7} h^{-1}. Therefore reliability tests can only be conducted in a reasonable time through accelerated testing. In an accelerated test one usually characterizes the item in stress conditions that exceed the normal stress conditions in service. A typical test consists in characterizing a device at higher temperatures than those that will be encountered in the field. A higher temperature will then, for example, induce a larger mechanical stress due to the mismatch of the coefficients of thermal expansion of two bonded materials.[6,7] To achieve genuine acceleration, the applied stress should still be lower than the technological or specification limits and should not induce another failure mechanism or alter the mechanism for which one assesses the corresponding failure rate. An adequate acceleration model then allows extrapolating the test results to the expected field stress. Extrapolated predictions nevertheless need to be taken with specific care. A main obstacle is the extrapolation of results gained with single-stress tests under carefully controlled laboratory conditions to the field application, i.e. an uncontrolled environment with combined stresses and extrinsic influences.[8]

The physics-of-failure method attempts to describe the physical, chemical or other process mechanism which leads to failure. A large number of failure mechanisms have been investigated and for some of those physical explanations for degradation effects have been proposed.[9–11] A top down physics-of-failure approach first identifies and quantifies the operational and environmental conditions acting on a component or system in terms of voltage, current, power, temperature, humidity, vibration, etc. The profound analysis of the thermal, electrical, mechanical, or chemical material properties allows describing the material behavior and the degradation in response to the applied stress and thus the potential failure mechanism can be identified and modeled.

Empirical failure rate prediction models represent a statistical interpretation of failure information collected from field data and from accelerated tests.[12–14] These models provide equations for the calculation of a predicted failure rate λ. Data input consists of component specific parameters such as component type, technology, packaging material etc., operational parameters, e.g. temperature, applied voltage etc., as well as of parameters that reflect the environmental impact. Such prediction models are based on the assumption that the failure rate of a large population of statistically identical and independent items can often be represented by the bathtub curve discussed in section 2.3. Empirical models describe the period during which the failure rate is assumed to be approximately constant. This assumption is not in contradiction to the fact that failures are caused by failure mechanisms. It rather expresses that none of the various underlying failure mechanisms is dominant for some period in time. During this equilibrium the random occurrence of failures can be statistically described by a constant rate.

The three methods introduced above come with their respective merits and shortcomings. They are therefore applied whenever their specific strength can be employed and the expenses can be justified.[15] Empirical prediction models however received lots of criticism, mainly because of the high sensitivity to assumptions about the input data and because of the varying results obtained with different prediction models.[16,17] In addition, making predictions on the basis of field data is not an easy task. First, the latter are very often very difficult to obtain from component manufacturers due to the proprietary nature and second, they can be incomplete as one did not necessarily analyze the failure mechanism involved neither the particular conditions in which the failure occurred.

Since no convincing alternative was found efforts were undertaken by the IEEE in order to give guidance on how to use prediction models in a comprehensive reliability assessment. This led to the development of the IEEE 1413 standard with the objective to identify required elements for an understandable, credible reliability prediction, which will provide the users of the prediction sufficient information to evaluate the effective use of the prediction results.[18,19] A reliability prediction according to this standard shall have sufficient information concerning inputs, assumptions, and uncertainty, such that the risk associated with using the prediction results would be understood.

Despite the debatable accuracy and other obvious disadvantages for the forecasting of field reliability, failure rate prediction models provide useful results for the comparative evaluation of intrinsic failure rates. The effort and cost for the generation of reliability test data is considered to be considerably higher. Even if the physics of failure approach would be suitable from a technical point of view it does not come into consideration due to the complexity, the associated cost and the weak applicability for complex systems. Empirical failure rate prediction models emerged from the investigations on failures and their statistical description for military electronics applications. The three most important models RIAC-HDBK-217Plus[12], Telcordia SR332[13] and IEC TR 62380[14] are briefly characterized in Table 1. Only models IEC TR 62380 and Telcordia SR 332 cover optical components to some level. However, the models are very basic and consist of a base failure rate and a multiplier accounting for the influence of temperature, mostly following an Arrhenius law.

TABLE 1. Main features of failure rate prediction models[12–14]

Model	RIAC-HDBK-217Plus	Telcordia SR 332	IEC TR 62380
Originator	US Dept. of Defence	Emerged from MIL-HDBK217	FranceTelecom RDF2000
International Standard	no	no	yes
Last update	2006	2006	2000
Base failure rates	yes	yes	yes
Temperat. cycles considered	yes	no	yes
Lifetime models	no	no	yes
Range of use	Military, all Industries	Mainly Telecom	Mainly Telecom

TABLE 2. Base failure rates at ambient temperature 40°C from Telcordia SR-332[13]

Device Type	λ_0 [FIT] in controlled (uncontrolled) environment
Fiber Optic Laser Module*	1000 (1500)
Fiber Optic LED Module*	240 (1100)
Fiber Optic Detect. Module*	500 (1400)
Fiber Optic Coupler/Splitter	180 (725)
WDM passive	550 (1500)
Optical Isolator	110
Optical Filter	1500
Single LED/LCD Display	3
Phototransistor, -detector	60, 15
Photodiode, -detector	15, 10
Light sensitive Resistor	20

*a module is a small packaged assembly including a laser diode/LED/detector and means for electrical connection and optical coupling

IEC TR 62380 in addition provides life expectancy models for various types of light emitting diodes (see for example Figure 3) and optocouplers. It is also the only model which provides a comprehensive consideration of temperature and temperature cycles while including the assessment of the lifetime of components with a lifetime limitation.

Table 2 and Table 3 summarize the base failure rates of the Telcordia SR-332 and the IEC TR 62380 model, respectively. Base failure rates of optical modules are provided in Table 4.

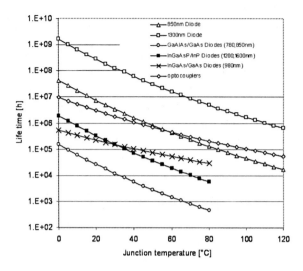

Figure 3. Life time of LEDs as a function of junction temperature from IEC TR 62380.[14]

TABLE 3. Base failure rates of passive and miscellaneous optic components from IEC TR 62380[14]

Component	λ_0 [FIT]
Attenuators: Bulk/fusion splice/fusion splice >10dB/pasted	2/2/10/10
Fusing-stretching couplers: 1 to 2/1 to n with n < 6	25/50
Integrated optical couplers 1 to n	60
Multiplexer and Demultiplexer	
Fusing-stretching 1 to 2/1 to n/micro-optic	25/50/60
Connectors, 1 optical contact	5
Jumper or optical cord, 2 optical contacts and fiber	10
Optical fiber (cable), per 100 km or section (any length)	500
Doped optical fiber, Si matrix (5 to 30 m)	1
LiNbO$_3$ modulator	1000
Isolator	10
Tunable filter	330*
Bragg array filter	15
Optical switch	200
VCSEL 840 nm	300*

* Values are given for a starting production

TABLE 4. Base failure rates of optical modules from IEC TR 62380[14]

Material	Module type	λ_0 [FIT]
GaAlAs/GaAs, 0.8 µm	Elementary emitter modules*	3000
InGaAs/InP 1.2–1.6 µm	Elementary emitter modules without (with) electronics*	40 (60)
InGaAs/InP 1.2–1.6 µm	Emitter/receiver module, with laser, PIN diode and electronics	80
InGaAs/InP 1.2–1.6 µm	Integrated modulator laser module	100
InGaAs/InP 1.48 µm	Pump laser module, p ≤ 250mW/ p > 250mW	200/350
InGaAs/GaAs 0.98 µm	Pump laser module	300
Si 0.7–1.1 µm/ InGaAs 1.2–1.6 µm	PIN diodes (PIN modules + electronics)	5/10 (30)
Si/Ge/InGaAs	APD diodes (APD modules + electronics)	20/40/80 (100)
Elementary fibered LED module (with driver)		100 (130)
Emitter/Receiver module, fibered LED + PIN diode (APD) + electronics		180 (200)
Optocouplers		10–100

* Generally an elementary laser module is made of a control photodiode, a laser diode and a coupling element.

2.5. CONCLUDING REMARKS ON RELIABILITY ENGINEERING

Ideally and to summarize, a complete reliability study of an optical component should involve the following activities.[2]

- Identify the service environment during normal service: temperature range, humidity range, chemical environment (including the species that could be released from the surrounding components and materials), solar or other radiation loads, mechanical stress, vibration levels and possible impacts, electromagnetic load, optical power levels in the component, etc.
- Define the component's operational conditions: optical power, current, voltage, frequency, etc.
- Define the failures and identify the failure mechanisms, i.e. state when a component fails and find the possible physical causes leading to this failure.
- Develop adequate ageing test methods, which simulate single or multiple service environmental conditions and operational conditions that can accelerate the ageing in a realistic manner.
- Gather laboratory data from ageing tests.
- From these develop lifetime and failure rate estimation/prediction theories, i.e. define models that allow to estimate the failure time

distribution and the failure probability for the dominating failure mechanisms during the specified lifetime and relevant to the service environment.

- Gather field data, i.e. collect the failure data from installed components including the failure time distribution and the analysis of the causes for the failures.
- Verify lifetime estimates against the field data and if necessary correct the lifetime estimation methods.

3. Reliability of Selected Optical Components

In the previous section we have seen that reliability engineering methods are well established. These well documented methods can be applied to the case of optical components. However to the knowledge of the authors and in spite of the existence of a number of reference documents that contain information on base failure rates of some types of optical components, there has not been any concerted or systematic approach so far for thoroughly assessing the failure rates and the reliability of all optical components and certainly not for optical fiber and waveguide sensors.

The situation is however not as dramatic as the paragraph above might suggest. Detailed studies have been conducted on a number of essential optical fiber sensor components, including not only the optical fibers themselves, but also fiber Bragg gratings. It is likely that the amount of relevant information will increase in the coming years. First statistically significant field data are appearing since a number of fiber sensors have been installed now for several years. For example the long-term durability of elongation sensors based on white light interferometry installed in concrete bridges was recently reported.[20] More than 100 sensors were observed over 6 years. Of these 100 sensors, 7 were damaged during installation, concrete pouring and bridge finishing. Five of the remaining sensors have stopped responding at subsequent times. The cause could not be determined exactly but in 3 cases there were indications that the measurement fiber failed or the passive part was damaged. From these observations one could estimate an 80% survival rate over 20 years.

3.1. MECHANICAL RELIABILITY OF OPTICAL FIBERS

The issue of mechanical reliability of optical fibers and optical cables is very important in the optical sensor field, for example in smart structures and structural health monitoring applications where fibers are embedded in materials or in the structures that need to be monitored and in which they are continuously experiencing fatigue and high stress levels.[21,22] Another example involves distributed sensors that rely on relatively long lengths of "standard" coated and cabled fiber or the lead fibers connecting the optical sensor probe to the data-acquisition equipment. We can also mention all the optical fibers interconnecting other types of fiber or waveguide sensors including multiplexed fiber Bragg gratings and Fabry-Perot cavity or interferometer type strain or displacement sensors.

The foundations of mechanical reliability of optical fibers were laid many years ago and a considerable amount of information on that subject is available in literature.[2] Optical fiber cables for the telecommunication industry are typically designed to have a lifetime exceeding 25 years. To quantify the mechanical reliability in a reasonable time one therefore has to conduct accelerated experiments at higher stress levels in the laboratory and extrapolate the results to the lower stress in-service conditions. A wealth of interesting information on how to deal with this has for example been published in a series of SPIE proceedings.[23–28] When browsing through these proceedings the reader will learn that different models have been proposed to estimate the lifetime of a fiber.

In this section we introduce the basics of optical fiber reliability as inspired by the Final Report of COST 246[2], Matthewson[29] and Baker[30] and the references therein through a physics of failure approach. The failure mechanism considered is fiber fracture due to weakening caused by stress-induced crack growth of flaws. Therefore to predict the time to failure for a given applied service stress or the maximum allowed service stress that guarantees a specified service life one conventionally uses fracture mechanics fatigue equations relying on power law crack growth kinetics. To do so one assumes that the crack growth rate v is proportional to a power of the stress intensity at the crack tip K_I

$$v = \frac{da}{dt} = AK_I^n \tag{8}$$

where a is the crack length, A and n are fit parameters. In fracture mechanics the stress intensity factor K_I is related to the perpendicularly applied stress σ and a shape parameter Y depending on the crack geometry through Eq. (9).

$$K_I = Y\sigma\sqrt{a} \tag{9}$$

Combining Eqs. (8)–(9) allows generating a differential equation that can be solved for a particular evolution of the applied load $\sigma(t)$. Fiber in a cable is for example considered to experience a long term loading event. In the static case $\sigma(t)$ is constant and one lets $\sigma(t) = \sigma_a$. One defines the initial or inert strength σ_i through

$$K_{IC} = Y\sigma_i\sqrt{a} \tag{10}$$

K_{IC} is the critical value of K_I, i.e. that value of the stress intensity at the crack tip for which catastrophic failure occurs. The initial strength decreases for larger crack sizes. Values of $K_{IC} = 8\cdot10^5$ Nm$^{-3/2}$ for silica glass and $\sigma_i = 13$ GPa for telecommunication fiber have been reported. Failure will occur when the decreasing strength reaches the applied stress.

In the static case one can find the time to failure (or fracture) t_f.

$$t_f = \frac{2}{(n-2)} \frac{\left(\sigma_i^{n-2} - \sigma_a^{n-2}\right)}{AY^2\sigma_a^n K_{IC}^{n-2}} \tag{11}$$

Equation (11) simplifies to

$$t_f = B\frac{\sigma_i^{n-2}}{\sigma_a^n} \tag{12}$$

with

$$B = \frac{2}{(n-2)} \frac{1}{AY^2 K_{IC}^{n-2}} \tag{13}$$

knowing that the applied stress is lower than the inert strength and that n is typically large for silica fibers.

For a static load the power law model thus links stress assisted fatigue to two environment dependent crack growth parameters: B which is the crack strength preservation parameter and a measure of the weakening effect of a particular applied stress history on the fiber, and n which is the fatigue constant and serves as a stress corrosion susceptibility parameter that measures how fast a crack grows when exposed to water and stress. Both values are necessary to calculate the lifetime under stress. n values ranging from 20 to 40 and B values from 10^{-8}GPa^{2}s to 0.1 GPa^{2}s for standard telecommunication fibers have been reported.

A similar reasoning can be followed for dynamic fatigue with a constant stress rate $\dot{\sigma}$ and applied load $\sigma(t) = \sigma't$.[29] In that case one can find the failure stress σ_f as

$$\sigma_f^{n+1} = (n+1)\sigma' B\sigma_i^{n-2} \tag{14}$$

Assessing the fiber lifetime in practice is relatively labor intensive. Laboratory testing typically consists in applying a well known stress to the fiber and measuring the time to failure in case of static load or stressing the fiber at constant strain rate and measuring the load at failure in the case of dynamic fatigue. This needs to be done repetitively on a statistically significant number of samples and preferably according to a standard method.[31-35] Experience has shown that for modern fibers 2% static strain is acceptable over civil engineering structure lifetimes, provided no additional surface cracks were initiated during installation.[36] In telecommunication applications, fiber failure due to stress and environmental corrosion assisted crack growth at fiber cable joints has typical failure rates from 0.5 to 2.3 FIT/km. This value is almost negligible compared to fiber damage due to cable dig-up, for which failure rates of 400 to 800 FIT/km have been reported.[37]

Fibers exhibit a modified mechanical strength following a zero-stress ageing process in corrosive conditions such as in humid conditions and in the presence of alkali and other ions. The chemistry at the fiber silica surface plays an essential role. It is important in all cases to limit mechanical damage to the fiber surface as this can result in the development of surface cracks. Additionally, water should be kept away from the fiber surface, for instance by hermetic coatings. However, many fiber sensors use modified fibers with locally removed coatings, local recoating or special types of coatings that allow a particular interaction for chemical sensing. It is known that stripping degrades the fiber strength. Removal of the coating without damage to the fiber is essential and one must make sure that the coating is completely removed. The presence of dust, humid air and chemicals during fiber treatment may cause additional strength degradation and the ageing may be accelerated by chemical residues. Additionally sensor fibers are often exposed to stress levels (in terms of temperature, mechanical solicitation and chemical aggressiveness) that may exceed those considered for standard telecommunication fiber cable. To the authors' knowledge there are only a few reports available in open literature on the mechanical reliability of such modified fibers in these very particular conditions.

3.2. FIBER BRAGG GRATINGS

The topic of optical fiber Bragg gratings (FBGs) and their characteristics is extensively covered in other chapters in this volume. From these descriptions it is obvious that FBGs are very popular and successful fiber sensor components. Besides the optical fiber itself, a fiber Bragg grating is probably the most extensively studied passive fiber sensor component from the reliability standpoint. FBGs show a complicated reliability picture due to the manifold of different failure mechanisms that may occur. First one has to cope with the degradation or decay of the refractive index pattern induced by thermal processes. It is obvious that an undesired change in the spectral characteristics of the FBG such as a shift of the center wavelength will lead to measurement drifts. A second possible failure stems from the reduction of the fiber strength following either the operator dependent handling of coating stripping and recoating or the UV laser illumination. A summary of these possible fiber Bragg grating failure mechanisms is schematically depicted in Figure 4.[38]Although not directly related to the component reliability itself, one should not forget that the accuracy and long-term stability of wavelength measurements is also essential to properly exploit the sensing characteristics of FBGs.[39,40]

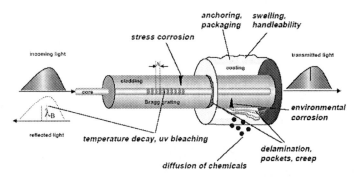

Figure 4. Illustration summarizing the different reliability issues pertaining to fiber Bragg gratings.[38]

The thermally induced decay of UV written FBGs in different types of optical fibers has been covered in open literature.[41–43] Erdogan[42] proposed a decay mechanism in which carriers excited during writing in non hydrogen-loaded Ge-doped fibers are trapped in a broad distribution of trap states and that the rate of thermal depopulation is an activated function of the trap depth. This model was found to be consistent with a power-law behavior. An important consequence of this mechanism is that the decay of the induced index change can be accelerated by increasing temperature. A continuous isochronal anneal method was also presented by Rathje[43] to analyze the thermal stability of UV gratings in both hydrogen loaded and unloaded fiber. Both reports use the concept of demarcation energy, which corresponds to the energy level separating populated and unpopulated defect trap states (Figure 5). Heating the FBG to a temperature T for a time t depopulates all the states below this demarcation energy E_d, given by

$$E_d = k_B T \, ln(v_0 t) \qquad (15)$$

where k_B is the Boltzmann constant and v_0 is a fitted value of the release rate (or attempt frequency), knowing that the release rate for a certain energy level E follows an Arrhenius law given by

$$v(E) = v_0 e^{-\frac{E}{k_B T}} \qquad (16)$$

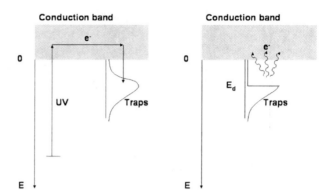

Figure 5. Physical mechanism underlying the thermal decay of fiber Bragg gratings. (a) UV excited electrons are trapped in a continuous distribution of traps, (b) thermal depopulation of the traps at a given temperature T for a certain time t – shallow traps ($E < E_d$) are being emptied and deeper traps ($E > E_d$) remain populated.

For FBGs the refractive index modulation is related to E_d through a so-called unique master curve, which plots the integrated coupling constant (*ICC*) or the normalized refractive index modulation against E_d, where the *ICC* is defined by Eq. (17).

$$ICC = \frac{\Delta n}{\Delta n_0} = \frac{arctan\left(\sqrt{R}\right)}{arctan\left(\sqrt{R_0}\right)} \qquad (17)$$

Δn represents the index modulation and R the reflectivity of the FBG. The subscripts 0 indicate the initial values. This curve can then be used to know at which temperature and during which time fiber Bragg gratings need to be pre-annealed to guarantee spectral stability within certain boundaries for specific operating conditions.[36] An example is given in Figure 6 which allowed to predict a reflectivity degradation of 2% maximum over 25 years and at 25°C for Bragg gratings written in hydrogen loaded germanium doped optical fibers. This however does not account for a possible decay of the mean refractive index of the gratings reflected by a shift of the Bragg wavelength. A second master curve obtained by representing the wavelength shift $\Delta\lambda$ versus the demarcation energy E_d is then necessary to predict wavelength changes with time and temperature. Proper pre-annealing of the gratings can guarantee a wavelength stability of 1 pm during 5 years at a temperature of 230°C, with a reflectivity decrease of less than 20%. Therefore, error-free operation can be assured through wavelength drift reduction to levels compatible with industry requirements.[44]

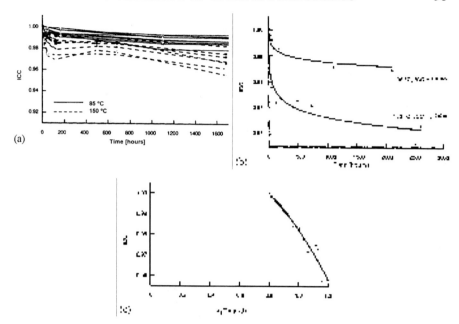

Figure 6. (a) ICC of a FBG written in hydrogen loaded germanium doped optical fiber as a function of time for two different temperatures, (b) Best power-law fit for the data in (a), and (c) Master curve fit for data in (b). (Adapted from Limberger[41]).

The mechanical reliability of FBGs has also been investigated. The fracture strength of gratings written with continuous wave UV sources has shown to be close to that of the pristine fiber, whereas FBGs fabricated using pulsed sources exhibit a decreased mechanical strength. As mentioned in section 3.1 the mechanical strength can be influenced by the stripping and recoating process. More detailed information on this issue has been reported by Limberger.[41] High-strength proof-testing was proposed as a way of ensuring a minimum strength. Lifetimes of the surviving samples in excess of 5 years under 0.5% static strain were shown to be achievable.[44] Upon UV writing of the grating in a locally stripped optical fiber recoating is necessary to ensure a minimum level of mechanical strength. One nevertheless has to be careful with coatings on FBGs. Besides the fact that they might introduce mechanical stress, the effect of environmental parameters on the coating itself might influence the response of the FBG. This has for instance been evidenced by Kronenberg[48] for polyimide coated gratings exposed to different levels of relative humidity. The temperature and relative humidity sensitivities were shown to increase with the coating thickness, as depicted in Figure 7.

In the previous paragraphs we have seen that methods exist to predict the lifetime of FBGs in particular conditions. A natural question then rises about the usefulness of that information when it comes to a multitude of grating sensors installed for monitoring large structures such as bridges or dams. Advanced reliability or sensitivity analysis methods relying on so-called Reliability Block Diagrams (RBDs) can then be applied. This allows deriving sensor network system failure rates or MTTF values depending on individual component failure rates as well as evaluating the effect of redundancy. Examples have been described by Held[44] and Brönnimann.[45,46]

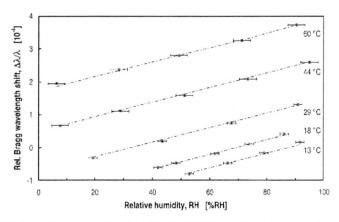

Figure 7. Relative Bragg wavelength shift of a FBG in a polyimide recoated optical fiber as a function of relative humidity for several temperatures.[48]

3.3. RELIABILITY OF MICRO-OPTICAL ELEMENTS

One of the major criticisms on optical fiber sensors is the relatively high cost of the associated read-out and analysis equipment. Efforts have there-fore been undertaken to decrease that cost, for example by introducing functional miniaturized or micro-optical devices that could easily be interfaced with or coupled to an optical fiber or waveguide and that could be fabricated at a relatively low cost. These miniature devices can then become an integral part of the sensor system and can provide – possibly very locally – processing and analysis of a transducer signal. On the other hand the micro-optical devices can act as sensor probes themselves. As an example of such a micro-optical sensor component we can mention a miniaturized optical spectrum analyzer.[49]

A current trend in the fabrication of such micro-optical devices consists in using a polymer as base material. Plastic materials indeed come with a number of important advantages such as (1) mechanical robustness com-pared with silicon and other semiconductor materials, (2) potentially lower preparation and processing cost, (3) low temperature processing and enhanced biocompatibility, (4) availability of flexible processing methods such as molding, embossing, melt processing and imprinting. Another advantage is the choice in polymer materials with application dedicated material properties. A variety of plastics have already been used for micro-optical applications, including photodefinable polyimide (e.g. Kapton), photoresist, SU-8 resist, silicone elastomer, PolyMethyl MethAcrylate (PMMA), PolyCarbonate (PC), Teflon, liquid crystal polymer and bio-degradable polymer.[50]

Moreover because of their low cost, low weight, mechanical flexibility and owing to the additional fabrication techniques that can be applied to the machining and bonding of polymeric materials when compared to the more traditional glass or silicon, polymers are often used as substrate material in micro-optical devices to carry optically functional coatings. Thin films of metal deposited on micro-structured polymer substrates for example are currently used to implement mirrors, waveguides and other optical elements.[51–53] In such applications typical thin film dimensions are a thickness of 50 to 500 nm and a grain size of several tens of nanometers to achieve an acceptable optical quality. The adhesion of inorganic films on the smooth and chemically inert surfaces of injection molded polymeric parts is usually poor. One therefore also applies an intermediate chromium adhesion layer.[54]

Many reliability problems occur when assembling materials with different characteristics. This is also the case here. The stability of the couple metal-film/polymer-substrate, in particular with respect to thermal cycling, and a good understanding of the related damage mechanisms are essential for the prediction and the improvement of the reliability of such systems. High temperatures appearing during soldering and bonding processes for packaging cause considerable thermo-mechanical stress. The latter will also appear during the service life of the component. Due to the difference in thermal expansion coefficients $\Delta\alpha$ between film and substrate, a temperature change ΔT indeed leads to a mismatch in elastic deformation $\Delta\varepsilon$ between the film and the substrate as given by Eq. (18).

$$\Delta\varepsilon = \Delta\alpha\Delta T \qquad (18)$$

This difference in deformation, which must be accommodated for the largest part by the film due to the much smaller film thickness compared to the substrate thickness, results in high residual stresses.

The relaxation of the elastic energy stored at the interface leads to either deformation of the substrate which curves, or film damage through roughening, delamination or cracking (Figure 8).[55–58] A change of the film surface morphology allowing stress relaxation has also been observed with a regular wave-like pattern appearing on the originally smooth surface. This phenomenon has been explained by mechanical buckling of the coating under compressive force.[59] Moreover fatigue of thin film on substrate systems results in extrusions and voids at the film/substrate interface underneath the extrusions depending on the film thickness and grain size. These voids extend from the interface towards the surface and crack-like features extend from the voids towards the surface. A number of authors report that there are no intrusions or cracks extending from the surface into the films. This fatigue damage is explained by the annihilation of edge dislocations, leading to a supersaturation of vacancies and to the observed voiding.[60,61]

It can be argued that at elevated temperatures diffusion driven damage mechanisms are accelerated. Furthermore during a thermal cycle the film is in tension at high temperature while being in compression at low temperature. Due to this asymmetry damage related to diffusion driven voiding and cracking occurring at high temperature may not be reversed upon cooling.

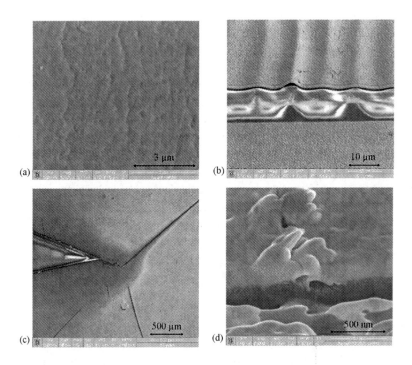

Figure 8. SEM micrographs of different types of damage observed in a micro-optical element (400 nm thick gold film on 1.28 mm thick PMMA substrate) after a thermal cycling according to the ISO 9022-2 norm: (a) roughening, (b) delamination, (c) cracking of the film, and (d) voids underneath extrusions.[58]

All the phenomena and mechanisms described above lead to a loss of the optical quality of the metal film and their onset should thus be considered as a failure parameter. The reliability assessment of micro-optical components is for example specified in the ISO 9022-2 norm[62] which relies on only 5 thermal cycles between –40 to +55°C. This test covers only the low cycle fatigue failure behavior of the micro-optical components under extreme conditions and is not adapted to the prediction of the long-term reliability of components, i.e. when several thousand temperature cycles of smaller amplitude need to be considered during the service life of the component.

4. Standardization Aspects

Whereas we already stated that the principles of reliability engineering can equally well be applied to components for fiber optic sensors, one has to realize that the implementation technologies and the market characteristics for fiber sensors are considerably different from the optical communications sector.[63,64] The devices in both applications have similar functions, but the demands put on their performance are generally quite different. Most of the efforts on optical component reliability have been conducted to serve the needs of the optical telecommunication industry. Optical communications standards are well established in contrast to those for the fragmented sensor market which typically delivers products dedicated to particular niche applications. This creates hurdles on the way of optical fiber or waveguide sensors to be accepted and applied. Compared to telecommunications the volume is low. Proprietary issues and custom specifications have kept the price of fiber optic sensors high. Many fiber optic sensors are commercialized by smaller companies that are not necessarily prepared to bear extended quality assurance or reliability programs. Only in very rare cases end-users are willing to pay for additional reliability assurance studies. Consequently the manufacturer will rather be inclined to qualify the complete sensor probe to a sequence of acceptance tests to provide assurance of initial performance. The reliability in long-term service is however not guaranteed by such acceptance tests.

To summarize one could say that there is a general lack of manufacturing infrastructure and of standards for packaging and reliability of optical fiber sensors. At the moment of writing this chapter a kind of consensus is rising on the necessity to encourage the introduction of standards for sensors and to take action for a better industry coordination. The fiber sensor market and the fiber sensor customer confidence are indeed likely to benefit from well-defined standard guidelines that would allow to compare the capabilities of different sensor types based on established standard procedures.[65]

Krohn[66] reported for example that the Fiber Optic Sensor Consortium has developed a preliminary standard for Bragg grating sensors, a Bragg grating sensor interrogator and an interferometric interrogator. The ASTM Subcommittee E13.09 on Fiber Optics, Waveguides, and Optical Sensors focuses on the formation and development of methods for testing, specifying, and using fiber optics, integrated optics, and other advanced optical and guided wave techniques for chemical sensing. The Petrotechnical Open Standards Consortium has drafted a communication or data transfer standard for distributed fiber temperature sensors referred to as Distributed Temperature Survey (DTS).[67]

Guidelines are also being defined by technical societies such as ISHMII (International Society for Structural Health Monitoring of Intelligent Infrastructure) in which fiber optic strain gauges are explicitly considered.[68] In 2005 the International Union of Laboratories and Experts in Construction Materials, Systems and Structures (RILEM) also set up a Technical Committee on optical fiber sensors. A COST Action 299: Optical Fibers for New Challenges Facing the Information Society (FIDES) in which a study group deals with fiber sensor guidelines was started in 2006. A first edition of IEC standard IEC 61757 has been published as well. It covers optical fibers components and subassemblies as they pertain specifically to sensing applications, in those aspects not already addressed by previous or concurrent standardization efforts. The standard defines, classifies and provides the framework for specifying fiber optic sensors and their specific components and subassemblies for fiber optic sensors.[69] IEEE also drafted a standard for single-axis interferometric fiber optic gyroscopes.[70]

5. Concluding Remarks

Reliability is a key topic for optical components and systems, for example when they are implemented as non-maintainable parts in infrastructure systems with long useful life. Models for life expectancy and failure rates are essential tools to assess the reliability of these components. In this respect reliability assessment methods developed during the past decades in the fields of mechanics and electronics can be applied to optical components. Such models ideally base on the characterization of the physical failure mechanisms supported by statistically firm (accelerated) testing which takes into account the relevant environmental and operational conditions. Some recommendations applicable to optical telecommunication components can be transferred to the sensor field but many problems particular to fiber optic sensing have to be addressed individually and guidelines still have to be established.

In addition, novel photonic technologies are being introduced at a substantial pace, while an extended range of application fields are being targeted. These technologies involve novel materials, encompass micro- and nano-photonics, rely on micro-opto-electro-mechanical systems, etc. Examples of new applications can be found among others in space, in aeronautics, in the biomedical field, in automotive industry and in nuclear facilities. Therefore reliability modeling and prediction of failure rates is likely to remain a very multidisciplinary and perpetual process that must continuously be adapted to the most recent technological developments and to increasingly challenging service environments.

Acknowledgements

The authors are very grateful to COST – the European Cooperation in the field of Scientific and Technical Research and in particular to COST Actions 246, 270 and 299 for providing a wealth of information presented in this chapter. The authors also acknowledge the European Commission Framework Program 6 Network of Excellence on Micro-Optics – NEMO. Many thoughts brought forward in the text have been shared during discussions with many people in the optical fiber sensor field, too many to mention here. Particular words of thanks go to Dr. Hans Limberger (EPFL, Switzerland), Dr. Wolfgang Habel (BAM, Germany), Dr. Rolf Brönnimann and Dr. Philipp M. Nellen (EMPA, Switzerland), and Prof. H. Thienpont (Vrije Universiteit Brussel, Belgium).

References

1. A. Birolini, *Reliability Engineering – Theory and Practice*, (Springer-Verlag, Berlin Heidelberg, 2004).
2. *Reliability of Optical Fibres and Components – Final Report of COST 246*, edited by T. Volotinen, W. Griffioen, M. Gadonna and H. Limberger, (Springer-Verlag, London, 1999).
3. M. Fukuda, *Reliability and Degradation of Semiconductor Lasers and LEDs*, (Artech House, London-Boston, 1991).
4. O. Ueda, Relaibility and Degradation of III-V Optical Devices, (Artech House, Boston-London, 1996).
5. C. Staveley, Applications of Optical Fibre Sensors to Structural Health Monitoring, Optimisation and Life-cycle Cost Control for Oil and Gas Infrastructures, *Business Briefing: Exploration & Production: The Oil & Gas Review 2004*; http://www.bbriefings.com.
6. H. Caruso and A. Dasgupta, A fundamental overview of accelerated testing, *Proceedings Ann. Reliability and Maintainability Symposium*, 389–393 (1998).

7. W. Nelson, *Accelerated Testing*, (Wiley Interscience, 1990).

8. Q.M. Meeker and L.A Escobar, Pitfalls of Accelerated Testing, *IEEE Transactions on Reliability* 47, 114–118 (1998).

9. E.A. Amerasekera and F.N. Najm, *Failure Mechanisms in Semiconductor Devices – Second Edition*, (John Wiley & Sons Ltd., Chichester, 1997).

10. M. Ohring, *Reliability and Failure of Electronic Materials and Devices*, (Academic Press, San Diego, 1998).

11. *Microelectronic Failure Analysis, Desk Reference – Fourth Edition*, (ASM International, 1999).

12. *HDBK-PLUS-WD – Handbook of 217Plus™ Reliability Prediction Models*, (RIAC, Utica NY, 2006).

13. *Reliability Prediction Procedure for Electronic Equipment*, Special Report Telcordia SR-332 Issue 2 (Telcordia, 2006).

14. *Reliability Data Handbook*, A universal model for reliability prediction of electronics components, PCBs and equipment, Technical Report IEC TR 62380, First Edition, (IEC, Geneva, 2004).

15. B. Foucher, J. Bouillie, B. Meslet and D. Das, A review of reliability prediction methods for electronic devices, *Microelectronics Reliability* 42, 1155–1162 (2002).

16. M. Pecht, Why the traditional reliability prediction models do not work – is there an alternative?, *Electronics Cooling* 2, (January 1996).

17. M. Pecht and W. Kang, A Critique of MIL-HDBK-217E Reliability Prediction Methods, *IEEE Transactions on Reliability* 37, 453–457 (1988).

18. IEEE Standard Methodology for Reliability Prediction and Assessment for Electronic Systems and Equipment, IEEE Std. 1413-1998 (IEEE, 1998).

19. M. Pecht, D. Das and A. Ramakrishnan, The IEEE standards on reliability program and reliability prediction methods for electronic equipment, *Microelectronics Reliability* 42, 1259–1266 (2002).

20. D. Inaudi, Long-term reliability testing of packaged strain sensors, *SPIE Proceedings* 5758, 405–408 (2005).

21. Z. Xu, A. Bassam, H. Jia, A. Tennant and F. Ansari, Fiber optic sensor reliability issues in structural health monitoring, *SPIE Proceedings* 5758, 390–404 (2005).

22. E. Udd, M. Winz, S. Kreger and D. Heider, *SPIE Proceedings* 5758, 409–416 (2005).

23. Optical Fiber Reliability and Testing, *SPIE Proceedings* 3848, edited by J. Matthewson (1999).

24. Optical Fiber and Fiber Component Mechanical Reliability and Testing, *SPIE Proceedings* 4215, edited by M.J. Matthewson (2001).

25. Optical Fiber and Fiber Component Mechanical Reliability and Testing II, *SPIE Proceedings* 4639, edited by M.J. Matthewson and C.R. Kurkjian (2002).

26. Reliability of Optical Fiber Components, Devices, Systems, and Networks, *SPIE Proceedings* 4940, edited by H.G. Limberger and M.J. Matthewson (2003).

27. Reliability of Optical Fiber Components, Devices, Systems, and Networks II, *SPIE Proceedings* 5465, edited by H.G. Limberger and M.J. Matthewson (2004).

28. Reliability of Optical Fiber Components, Devices, Systems and Networks III, *SPIE Proceedings* 6193, edited by H.G. Limberger and M.J. Matthewson (2006).

29. M.J. Matthewson, Strength Probability Time Diagrams using Power Law and Exponential Kinetics Models for Fatigue, *SPIE Proceedings* 6193, 619301 (2006).

30. L.K. Baker, Comparison of Mechanical Reliability Models for Optical Fibers, Corning White Paper WP5049 (2001).

31. IEC 60793-1-30 First Edition, Optical fibres – Part 1–30: Measurement methods and test procedures – Fibre proof test (IEC, 2001).

32. IEC 60793-1-31 First Edition, Optical fibres – Part 1–31: Measurement methods and test procedures – Tensile strength (IEC, 2001).

33. IEC 60793-1-32 First Edition, Optical fibres – Part 1–32: Measurement methods and test procedures – Coating strippability (IEC, 2001).

34. IEC 60793-1-33 First Edition, Optical fibres – Part 1–33: Measurement methods and test procedures – Stress corrosion susceptibility (IEC, 2001).

35. TIA/EIA-455-28-C FOTP-28 - Measuring Dynamic Strength and Fatigue Parameters of Optical Fibers by Tension (TIA/EAI, 2004)

36. U. Sennhauser and Ph.M. Nellen, in: *Trends in Optical Nondestructive Testing and Inspection*, edited by P.K. Rastogi and D. Inaudi (Elsevier Science Ltd. Oxford, 2000), pp. 473–485.

37. T. Volotinen, C. Kurkjian and A. Opacic, in: *Reliability of Optical Fibres and Components – Final Report of COST 246*, edited by T. Volotinen, W. Griffioen, M. Gadonna and H. Limberger (Springer-Verlag, London, 1999), pp. 37–77.

38. P. Mauron, *Reliability and lifetime of optical fibres and fibre Bragg gratings for metrology and telecommunications*, Thèse n° 2339 (Ecole Polytechnique Fédérale de Lausanne, Switzerland, 2001).

39. S.D. Dyer, P.A. Williams, R.J. Espejo, J.D. Kofler and S.M. Etzel, Key metrology considerations for fiber Bragg grating sensors, *SPIE Proceedings* 5384, pp. 181–189 (2004).

40. A. Fernandez Fernandez, A. Gusarov, F. Berghmans, K. Kalli, V. Polo, H. Limberger, M. Beukema and Ph. Nellen, Round-robin for fiber Bragg grating metrology during COST270 action, SPIE Proceedings *5465*, pp. 210–216 (2004).

41. H.G. Limberger, in: *Reliability of Optical Fibres and Components – Final Report of COST 246*, edited by T. Volotinen, W. Griffioen, M. Gadonna and H. Limberger (Springer-Verlag, London, 1999), pp. 326–384.

42. T. Erdogan, V. Mizrahi, P.J. Lemaire and D. Monroe, Decay of ultraviolet-induced fiber Bragg gratings, *Journal of Applied Physics* 76, pp. 73–80 (1994).

43. J. Rathje, M. Kristensen and J.E. Pedersen, Continuous anneal method for characterizing the thermal stability of ultraviolet Bragg gratings, *Journal of Applied Physics* 88, pp. 1050–1055 (2000).

44. M. Held, R. Brönnimann, Ph. M. Nellen and L. Zhou, Relaibility engineering – basics and applications for optoelectronic components and systems, *SPIE Proceedings* 6188, 618815 (2006).

45. R. Brönnimann, Ph.M. Nellen and U. Sennhauser, Reliability monitoring of CFRP structural elements in bridges with fiber optical Bragg grating sensors, *Journal of Intelligent Material Systems and Structures* 10, 322–329 (1999).

46. R. Brönnimann, M. Held and Ph.M. Nellen, Reliability, Standardization and Validation of Optical Fiber Sensors, *18th International Optical Fiber Sensors Conference Technical Digest*, ThD1 (Optical Society of America, Washington DC, 2006).

47. Ph.M. Nellen, P. Mauron, A. Frank, U. Sennhauser, K. Bohnert, P. Pequinot, P. Bodor and H. Brändle, Reliability of fiber Bragg grating based sensors for downhole applications, *Sensors and Actuators A 103*, 364–376 (2003).

48. P. Kronenberg, P.K. Rastogi, P. Giaccari and H.G. Limberger, Relative humidity sensor with optical fiber Bragg gratings, *Optics Letters* 27, 1385–1387 (2002).

49. T. Mappes, S. Achenbach, A. Last, J. Mohr and R. Truckenmüller, Evaluation of optical qualities of a LIGA-spectrometer in SU-8, *Microsystem Technologies* 10, 560–563 (2004).

50. C. Liu, J. Chen, J. Engel, J. Zou, X. Wang, Z. Fan, K. Ryu, K. Shaikh and D. Bullen, Polymer micromachining and applications in sensors, microfluidics, and nanotechology,

The 226th National Meeting of the American Chemical Society (ACS), New York, USA, 11–17 Sept 2003. Available in *Polymer Reprints* 44, (2003).

51. N. Bowden, S. Brittain, A.G. Evans, J.W. Hutchinson and G.M. Whitesides, Spontaneous formation of ordered structures in thin films of metals supported on an elastomeric polymer, *Nature* 393, 146–149 (1998).

52. H. Kupfer and G.K. Wolf, Plasma and ion beam assisted metallization of polymers and their application, *Nuclear Instruments and Methods in Physics Research B* 166–167, 722–731 (2000).

53. A. Last, H. Hein and J. Mohr, Shape deviations in masks for optical structures produced by electron beam lithography, *Microsystem Technologies* 10, 527–530 (2004).

54. C. Malins, T.G. Harvey, P. Summersgill, P. Fieldens and N.J. Goddard, Embossed polymer leaky waveguide devices for spectroscopic analysis, *Analyst* 126, 1293–1297 (2001).

55. W.D. Nix, Mechanical properties of thin films, *Metallurgical and Materials Transactions* 20A, 2217–2245 (1989).

56. Y.C. Zhou and T. Hashida, Thermal fatigue failure induced by delamination in thermal barrier coating, *International Journal of Fatigue* 24, 407–417 (2002).

57. M. Huang, Z. Suo, Q. Ma and H. Fujimoto, Thin film cracking and ratcheting caused by temperature cycling, *Journal of Materials Research* 15, 1239–1242 (2000).

58. S. Eve, N. Huber, O. Kraft, A. Last, D. Rabus and M. Schlagenhof, Development and validation of an experimental setup for the biaxial fatigue testing of metal thin films, *Review of Scientific Instruments* 77 (2006).

59. A.L. Volynskii, A. Bazhenov, O.V. Lebedeva and N.F. Bakeev, Mechanical buckling instability of thin coatings deposited on soft polymer substrates, *Journal of Materials Science* 35, 547–554 (2000).

60. R. Schwaiger, G. Dehm and O. Kraft, Cyclic deformation of polycristalline Cu films, *Philosophical Magazine* 83, 693–710 (2003).

61. O. Kraft, P. Wellner, M. Hommel, R. Schwaiger and E. Arzt, Fatigue behavior of polycristalline thin copper films, *Zeitschrift für Metallkunde* 93, 392–400 (2002).

62. ISO 9022-2:2002 norm, Optics and optical instruments – Environmental test methods – Part 2: Cold, heat and humidity (2002).

63. B. Culshaw, Optical Fiber Sensor Technologies: Opportunities and – Perhaps – Pitfalls, *Journal of Lightwave Technology* 22, 39–50 (2004).

64. B. Culshaw and W. Habel, Fibre Sensing: Specifying Components and Systems, *Technical Digest: Symposium on Optical Fiber Measurements 2004, NIST Special Publication* 1024, 179–189 (2004).

65. W. Habel, Requirements and Methods to Establish Confidence in Use of Fibre Optic Sensors, *18th International Optical Fiber Sensors Conference Technical Digest*, MD4 (Optical Society of America, Washington DC, 2006).

66. D. Krohn, Market Opportunities and Standards Activities for Optical Fiber Sensors, *18th International Optical Fiber Sensors Conference Technical Digest*, FB1 (Optical Society of America, Washington DC, 2006).

67. DTS Draft Data Schema Specification; http://www.witsml.org

68. Guidelines for Structural Health Monitoring, Design Manual No. 2 (ISIS Canada, 2001).

69. IEC 61757-1 First Edition, Fibre optic sensors – Part 1: Generic specification (IEC, 1998)

70. IEEE 952-1997, IEEE Standard Specification Format Guide and Test Procedure for Single-Axis Interferometric Fiber Optic Gyros (IEEE, 1997)

DEVELOPMENT OF THE DISTRIBUTED BRILLOUIN SENSORS FOR HEALTH MONITORING OF CIVIL STRUCTURES

XIAOYI BAO* AND LIANG CHEN
*Physics Department, University of Ottawa, Ottawa, Ontario
Canada K1N 6N5*

Abstract. The progress of the distributed fiber sensors based on Brillouin scattering has been reviewed, the system limitation and improvement of spatial resolution using the different signal processing schemes and the simultaneous monitoring of temperature and strain have been summarized. The applications of distributed Brillouin sensors for structural health monitoring are provided.

Keywords: Fiber optical sensors, Brillouin scattering, structural health monitoring, the temperature and strain sensors

1. Introduction of Structural Health Monitoring and Distributed Sensors

Structural Health Monitoring (SHM) has been used to identify early signs of potential problems of civil structures to prevent disasters, and conduct needed repairs at the appropriate time to avoid unnecessary costs and reduce economic burden. Thus it is important to have accurate and real time monitoring on the safety assessment of civil structures, such as bridges, dams and pipelines. The key is to prevent the potential disasters. Currently such evaluations are carried out by engineers trained in visual inspection, which sometimes can be inaccurate due to the personal experience differences on the safety condition assessment generated by this practice. To increase the inspection efficiency and accuracy, various sensors have been developed and being demonstrated in the field. Among many sensors being used for civil structural monitoring, fiber optic sensors are one of the most

---

*To whom correspondence should be addressed.

W.J. Bock et al. (eds.), Optical Waveguide Sensing and Imaging, 101–125.
© 2008 *Springer.*

promising candidates due to their features of durability, stability, small sizes
and insensitivity to external perturbations, which makes them ideal for the
long-term health assessment of civil structures (Tennyson et al., 2000;
Culshaw, 2004).

Most of the traditional sensors used in civil engineering applications are
strain gauges, or fibre optical sensors, such as Bragg grating, long gauge or
Fabry Perot type of sensors. They are point sensors that provide local stress
information at pre-determined specific points. However, for SHM the cracks
and deformation locations are often unknown in advance, and thus it will be
difficult for point sensors to be placed at the right points to be close to
"potential" crack, deformation or buckling locations. Because of the single
point detection, it is unlikely to accurately correlate the strains at the different
locations to the status of the structures, which is directly related to SHM in
which the structural strain monitoring is required to assess the safety of the
structures. This could be achieved with the distributed fiber sensors using
optical fibers to cover the large areas of the civil structures and to access the
safety and status of the structures. This kind of sensor has the advantages of a
long sensing range and the capability of providing the strain or temperature at
every spatial resolution over the entire sensing fiber imbedded in or attached
on the structures by using the fiber itself as the sensing medium.

Spatial information along the length of the fiber can be obtained through
optical time domain reflectometry (OTDR), i.e. measuring pulse propagation times for light traveling in the fiber. The length covered by the
optical pulse is the spatial resolution. This allows for continuous monitoring
of the structural strain in a distributed fashion, which can be obtained by the
distributed Brillouin scattering or Rayleigh scattering. The most commonly
used measurement technique is the Brillouin scattering based distributed
sensing as it provides long sensing length for the measurement of the
temperature or strains.

2.   Brillouin Scattering and Distributed Brillouin Sensor

The distributed fiber sensor based on Brillouin scattering exploits the
interaction of light with acoustic phonons propagating in the fiber core. The
Brillouin scattering light is in the backward direction with a frequency shift
proportional to the local velocity of the acoustic phonons (also called
acoustic waves), which depends on the local density and tension of the glass
and thus the material temperature and strain. The Brillouin frequency shift,
V_B is in the order of 9–13 GHz for the radiation wavelength of 1.3–1.6 μm
in standard single mode communication fibers, and it is given by $V_B = \frac{2nV_a}{\lambda}$,
where V_a is the velocity of sound in glass, n is effective refractive index of
the optical fiber, and λ is the free-space wavelength (Boyd, 2003). The

sensing capability of this scattering phenomenon arises from the measurement of distributed Brillouin frequency shift dependence on both strain and temperature.

The concept of using Brillouin scattering for fiber optic sensing was first proposed in 1989 (Horiguchi and Tateda, 1989) and it was termed Brillouin optical time domain analysis (BOTDA). This approach involved launching a short pump pulse into one end of the test fiber and a CW (continuous wave) probe beam into the other end. The frequency difference between the two lasers could be set to a particular value corresponding to the Brillouin frequency of the optical fibers, and the CW probe would experience gain at locations along the fiber. The gain as a function of position along the fiber could thus be determined by the time dependence of the detected CW light. By measuring the time dependent CW signal over a wide range of frequency differences between the pump and probe, the Brillouin frequency for each fiber location could be determined. This allowed for the strain or temperature distribution along the entire fiber length to be established.

The first strain distribution measurement was reported on submarine cables (Tateda et al., 1990) using distributed fiber sensors based on Brillouin scattering. The reported strain distribution was obtained over a 1.3-km cable (Horiguchi et al., 1989). Later, another strain test on the bent slot-type optical cables was reported (Horiguchi et al., 1992). Temperature measurement using Brillouin scattering was proposed in 1989 (Culverhouse et al., 1989) utilizing the linear relationship between Brillouin frequency shift and temperature in a single mode fiber and measured by a Fabry-Perot interferometer. Then, a distributed temperature measurement on a 1.2 km single mode fiber with a 3°C temperature resolution and a 100 m spatial resolution was demonstrated with a BOTDA system (Kurashima et al., 1990), as shown in Figure 1.

The next development in BOTDA was the use of Brillouin loss rather than Brillouin gain (Bao, 1993) in order to increase the sensing length. Instead of the frequency of the pump (continuous wave) being greater than the frequency of the probe (pulsed beam), the opposite was implemented. The CW probe therefore experienced loss, rather than gain, at locations along the fiber at which the frequency difference between the lasers matched the local Brillouin frequency of the fiber. As the pump pulse is not depleted in this case, longer sensing lengths of 50 kilometers for 10 m spatial resolution was reported using the Brillouin loss mechanism (Bao et al., 1995). The longest reported distributed sensor length using spontaneous Brillouin scattering is 57 km with a spatial resolution of 20 m (Maughan et al., 2001). More recently, sensing length over 100 km has been demonstrated by amplifying the signal in the fiber (Alahbabi et al., 2006).

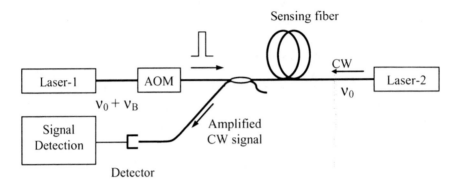

Figure 1. Gain based BOTDA configuration.

Another development using Brillouin scattering is a single-ended BOTDA system (Fellay, 1997). An amplitude modulator generates frequency-shifted sidebands that can serve as a probe. By eliminating the second laser, the cost of the system is reduced. However, the added costs of the frequency modulator, drivers and erbium doped fiber amplifier result in the total cost savings being minimal. This system has been used for temperature monitoring in the field. Fiber was embedded in the Luzzone dam in the Swiss Alps for monitoring of the temperature of the concrete during curing (Thévenaz et al., 1998).

3. Spatial Resolution Improvement to Sub-meters (<phonon lifetime)

Thus, Brillouin scattering based systems were well suited for long range sensing, such as communication cable strain or temperature monitoring. On the other hand, strain monitoring for civil engineering and aerospace structures required much shorter spatial resolutions than were achievable, as well as relatively shorter sensing length than the regular cable communication fibers. Recent research effort has been directed towards reducing the spatial resolution to the point where it is practical for structural monitoring applications. Two solutions were successfully proposed and demonstrated in the past few years to achieve the sub-meter spatial resolution.

The first solution is to avoid the spectral spreading of the Brillouin spectrum due to the broadband pulse signal by generating a pre-pumping of the acoustic wave using weak CW leakage (DC level of the un-modulated pulse signal) of the pulse, and then making the interaction of pump and probe on this pre-pumped acoustic wave through the leakage of the pulse and CW pump (Bao et al., 1999; Kalosha et al., 2006). This solution combines the advantage of an ideally narrow Brillouin gain spectrum together with the high spatial resolution of very short pulses. Spatial resolutions down to 10 cm have been demonstrated using this scheme. This is remarkable, as it was believed that 1 m spatial resolution would be the best that could ever be achieved due to the finite lifetime of the acoustic phonons (Horiguchi et al., 1995). The most recent developments on this technique have been focused on structural health monitoring of various civil structures, which is outside the domain of fiber communications (Ohno et al., 2001).

The second solution is to generate frequency-dithered signals that propagate in opposite directions (Hotate and Hasegawa, 1998). If the dithering frequencies are identical, it can be demonstrated that the frequency difference between the 2 waves is constant only at periodic definite locations along the fiber. At these points the stimulated Brillouin interaction is very efficient and can be extremely localized. The position of the interaction point can be scanned along the fiber by varying the dithering frequency, although the scanning time can be very long (Hotate and Hasegawa, 2000), centimeter spatial resolution was demonstrated using this scheme (Hotate and Tanaka, 2002).

4. Simultaneous Temperature and Strain Sensing Using the Distributed Brillouin Sensors

One problem with the implementation of Brillouin scattering based sensing systems in the field is the sensitivity of the Brillouin frequency shift to both strain and temperature. This leads to ambiguity in the measurement, as one does not know whether the frequency shift was caused by the change of strain or temperature. In a laboratory environment the temperature is essentially constant and its effects can generally be neglected when measuring strain. In many field situations this is not the case.

An early solution to this problem proposed the use of two fibers placed adjacent to one another, in which one fiber was isolated from any strain effects (Bao et al., 1994). The isolated fiber would be used to monitor the temperature, while the other fiber would measure the effect of both strain and temperature. Recent innovative approaches to measuring the strain and temperature simultaneously have been developed. By combining the Landau-Placzek ratio with the frequency shift, the temperature and strain were determined simultaneously (in the spontaneous Brillouin scattering regime) at a spatial resolution of 40 m (Parker et al., 1998). Most recently, using the same principle, an improved system has been reported which achieved of a temperature resolution of $4°C$, a strain resolution of 290 micro-strains and a spatial resolution of 10 m for a sensing length of 15 km (Huai et al., 2000).

Centimeter resolution has been achieved with simultaneous temperature and strain sensing using Brillouin loss based distributed sensor with polarization maintaining fibers (PMF) and photonic crystal fibers (PCF).

The simultaneous temperature and strain sensing using the PMF can be realized with Stokes power (intensity) and Brillouin frequency. Unlike the single mode fiber where the intensity fluctuation is overwhelmed by the polarization mode dispersion induced polarization change in the fiber, the intensity fluctuation in PMF is only caused by the light source power fluctuations, if the temperature or strain is constant, provided the light is launched in one of the PMF axes. If the intensity change of the light source is negligible, the intensity change caused by the temperature and strain can be measured accurately (Yu et al., 2004). Furthermore when the frequency-stabilized lasers are used, the linewidth of the laser is very narrow and change slowly with time. Therefore the dependence of the Brillouin spectrum width on the temperature and strain can be measured accurately. Thus three parameters (Brillouin frequency, Stokes power and Brillouin spectral width) can be used for the temperature and strain measurement simultaneously at temperature accuracy of $< 1°C$ and strain accuracy of $< 20\mu\varepsilon$ (Bao et al., 2004).

In PCFs with solid silica core the guiding mechanism is the same as the conventional single mode fiber, except the effective cladding index is the average of the air and silica refractive indices. The solid silica core with a Ge-doped center region can increase the nonlinear refractive index of the core and create a smaller mode field diameter. As a result, the Brillouin spectrum of PCF shows multiple peaks with comparable intensities (Zou et al., 2003) with a main peak and several sub-resonance peaks due to guided acoustic modes. The temperature coefficient for the main resonance and a peak originating from a higher-order guided longitudinal acoustic mode in the PCF with a partially graded Ge-doped core were identified for simultaneous temperature and strain measurement using their Brillouin frequency shifts only. It allows high temperature and strain measurement accuracy with spatial resolution of 20 cm (Zou et al., 2004).

5. The Offset Locking of DFB Lasers and Bias Control of the Optical Modulator

5.1. OFFSET LOCKING THE PUMP AND PROBE WAVE AT BRILLOUIN FREQUENCY USING THE OPTICAL DELAY

One of the criteria in getting high quality performance of the distributed sensor system is to phase lock the pump and probe wave at the Brillouin frequency using a stabilized laser source. As the optical modulator is used to generate the optical pulse through phase modulation, it is very important to phase lock the pump and probe lasers. An offset locking technique (Doi et al., 2001) was used on two DFB lasers. This technique was used to stabilize the frequency of a millimeter-wave sub carrier, except a PID controller (hardwire) was employed to lock the beat frequency, which provides a fast response determined by the optical delay line (Chen et al., 2004). The frequency tuning is achieved by tuning the delay time of the optical delay line, which has no limitation for the frequency difference as shown in Figure 2. The only limitation to the tuning frequency range is the delay range. To keep the optical modulator at high extinction ratio, we used the lock-in amplifier to stabilize the bias drift at minimum DC level. To minimize the Brillouin spectrum distortion at the long fiber length (kilometers) the pump power was limited to less than 10 mW. This allows high signal to noise ratio of over 35 dB, which gives Brillouin frequency measurement accuracy of 0.5 MHz for 10 ns pulses over 2 km fiber length.

To reduce the polarization dependence of the Brillouin gain spectrum, we used polarization scrambling at a rate of 12 KHz. Temperature resolution of better than 1°C was demonstrated.

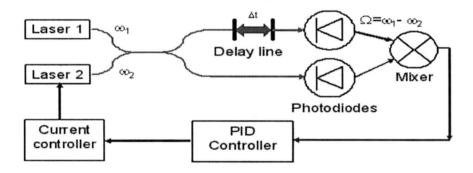

Figure 2. Laser frequency offset locking system.

5.2. EOM BIAS CONTROL USING LOCK-IN AMPLIFIER AND PID CONTROLLER

In order to produce the pulsed probe beam, an electro-optic modulator (EOM) is used to turn the probe beam on and off to acheive the spatial resolution for distributed sensors. The EOM can never block 100% of the light from the laser and so the probe beam will always have a DC component in addition to the pulse. In the sensor system, the DC component is sometimes beneficial (Afshar et al., 2003) and sometimes harmful to the sensor performance (Lecoeuche et al., 1999) depending on the sensing length and spatial resolution. For long sensing length (>10 km) with nanosecond pulses it is recommended to have the DC level (the un-modulated pulse portion) as low as possible, as the nonlinear Brillouin scattering would produce large DC background which leads to overshoot in the Brillouin gain/loss signal resulting a distorted Brillouin spectrum (Bao et al., 2006) and increase the uncertainty of the Brillouin frequency shift measurement. This leads to a power dependent acoustic wave relaxation time, which should be avoided in the distributed Brillouin sensors. On the other hand, a moderate DC level does enhance the sensor performance for centimetre spatial resolution (Kalosha et al., 2006), i.e. minimum detectable stress and temperature section.

In addition to the need of DC level control for sensor performance enhancement at centimetre spatial resolution based on Brillouin gain or loss, the bias of the EOM is drifting over time even when a stabilized voltage is applied. This means that the pulse energy fluctuates over time, causing SNR variation and higher measurement error.

There are two methods applied to EOM bias control (Snoddy et al., 2006): 1) to lock the pulse base using a lock-in amplifier as shown in Figure 3; 2) to lock the pulse base to a set value using a proportional-integral-derivative (PID) control algorithm implemented in software as shown in Figure 4.

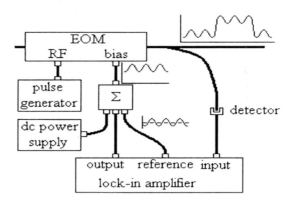

Figure 3. Setup of lock-in amplifier method of bias locking.

Each method has its own benefits and drawbacks depending on the type of applications and the requirements on the sensing length and spatial resolution. For large pulses and long fiber length it is desirable to provide Stokes pulses with the largest possible extinction ratio – in this case, the lock-in amplifier method of bias locking is clearly superior.

The lock-in amplifier method locks the bias voltage to the minimum of the EOM transfer curve; it is impossible to do so with the PID method

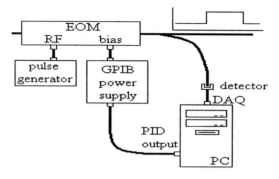

Figure 4. Setup of PID and computer control method for bias locking.

because it must operate on a region of the transfer curve increasing mono-tonically. For the cases that accuracy and spatial resolution of a centimeter are needed (1–10 ns), then a lower extinction ratio (10–20 dB) can be locked by the PID method. In applications that require large extinction ratio and long fiber length, the lock-in amplifier method has better performance.

6. Signal Processing for Improvement of the Spatial Resolution

In section 3 there are two successful approaches demonstrated for the sub-meter spatial resolution by modification of the light source; one requires narrow pulse with broadband spectrum, the other one requires long pro-cessing time and complicated frequency detection scheme. The spatial resolution of the distributed sensor is defined as the pulse width covered length. Here three signal processing methods are introduced to improve the spatial resolution to better than the pulse length by 1) identifying the stress boundary using compound spectral method; 2) second order derivative of frequency and location; and 3) introducing Rayleigh criteria to get sub-pulse length spatial resolution.

6.1. DE-CONVOLUTION OF COMPOUND BRILLOUIN SPECTRUM TO IMPROVE THE SPATIAL RESOLUTION

For a fixed pulse length, one can send pulses twice at different times with one pulse length delay for the second pulse, so that the Brillouin spectrum is de-convoluted in the same position twice to cover the half of the pulse length, which means the spatial resolution is improved by a factor of two (Brown, 1999). Similar idea can be used to get a factor of four improvement with ½ of the pulse length delay, so that in any given position, the Brillouin spectrum will be de-convoluted by four times.

Under uniform strain within the pulse length, the Brillouin loss spectrum is a single peak at a beat frequency corresponding to the strain and temperature in the fiber. The line width depends on the optical pulse length used. If the strain or temperature within the pulse width is non-uniform, the Brillouin loss profile contains multiple peaks. The individual components of the compound spectrum may be resolved using signal processing, if the first section of the strain or temperature is known.

Assuming the Brillouin gain coefficient is constant over the length of a fibre section of constant strain, the time-domain signal is a convolution of the shape of the section of uniform strain (rectangular) with the shape of the pulse (also approximately rectangular). Considering a fibre section shown in Figure 5a, as the pulse enters the section the signal increases to the point where the pulse is entirely within the section. Then the signal maintains at this level for as long as the pulse is within the section and starts to decrease

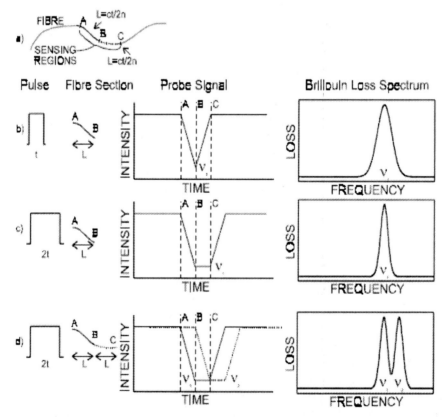

Figure 5. Time-domain waveforms and Brillouin loss spectra for various pulse widths. From Brown et al. (1999) ©IEEE.

as the pulse leaves the section. For a pulse the same width as the strained section, this results in a triangular waveform (Figure 5b). A pulse twice the width of the section will produce a trapezoidal shape in the time-domain waveform (Figure 5c). If there are two adjacent sections of different uniform strain (and thus different Brillouin frequency), the trapezoidal shape from the latter section will be partially overlapped by that of the earlier section. If the time-domain signal is measured at a point in this overlap, the frequency domain signal contains the spectrum of both sections (Figure 5d). A pulse width twice the length of a single fibre loop generates a spectrum containing two Brillouin lines corresponding to the strains of the two sections covered by the pulse. It is possible to use the compound spectra of N sections measured two at a time, together with a prior know-ledge of the signal from one of the sections to determine the Brillouin loss spectrum of the other N-1 sections.

6.2. THE SECOND-ORDER DERIVATIVE OF THE FREQUENCY AND POSITION TO IDENTIFY STRAIN CHANGE

Finding the boundary of two strained sections is a difficult task, especially for a large strain gradient or a small strain change. It involves multiple peaks (for large Brillouin frequency variation) or a broadened Brillouin profile (small Brillouin frequency variation). Because different strain com-ponents make up the Brillouin spectrum, there will be a transition between different strain components. On the other hand, each strain component belongs to a different location in the time domain Stokes signal being added within the same pulse length, this should show another transition in the position map as shown in Figure 6. Thus a mixed second-order partial

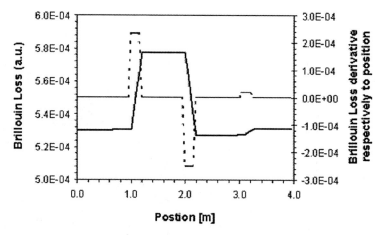

Figure 6. Brillouin loss (solid line) and its 1st order derivative (dashed line). From Yu et al. (2005). ©OSA

derivative of the Stokes intensity (Brillouin loss signal) with respect to frequency and location can be used to identify the boundary for different Brillouin frequencies induced by the tempertaure or strain changes. For instance, if a 2 ns pulse equivalent to 20 cm spatial resolution, is used for the distributed Brillouin sensor, 5 cm location accuracy could be obtained between two strain sections, which is limited by the readout resolution, i.e. the time resolution of the digitizer. This means the location accuracy is ¼ of the pulse length (Yu et al., 2005). So one can get four times improvement for the location accuracy without using the narrow pulse which could lead to broadband electronics and higher noise, as well as expensive electronics. The signal processing is done by the computer programming rather than hardware.

6.3. RAYLEIGH EQUIVALENT CRITERIA TO IDENTIFY THE STRAIN SECTION SHORTER THAN THE SPATIAL RESOLUTION

The Rayleigh criterion allows the determination of the smallest resolvable angular separation of two identical objects. It uses two distributions of equal intensity whose equations have the generic shape $I = \text{sinc}^2 \delta$ where δ also called phase, is the normalized distance between two objects. The criterion assumes that these two peaks can be resolved as soon as the maximum intensity of the first peak coincides with the first minimum of the second peak, which happens for $\delta = \pi$. The distance between these two objects is then the smallest resolvable distance. The intensity at the minimum distance between two objects is $8 / \pi^2$. This idea can be applied to the distributed Brillouin sensor in the Brillouin gain or loss spectrum to identify the two separated strain or temperature peaks. The intensity is the Lorentzian shaped Brillouin spectrum, and the phase is the Brillouin frequency. Figure 7 shows

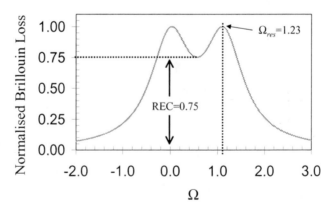

Figure 7. Definition of the Rayleigh Equivalent Criterion for simulated Brillouin loss spectrum with the fiber length of 1000 m, z = 0, Pp = 30 mW, Pcw = 5 mW, pulse length = 20 m. From Fabien Ravet' s PhD Thesis (2007).

Brillouin loss spectrum obtained by simulation of the phenomenological model (Ravet, 2005). The dip amplitude (minimum of the Brillouin loss spectrum comprised between the two peaks) is 0.75 corresponding to $\Omega_{Bs} = \Omega_{res} = 1.13$.

Here $\Omega_{Bs} = \frac{v_{Bs} - v_B}{\Delta v_B}$, and Ω_{res} represents the smallest resolvable frequency difference. Δv_B is the bandwidth of the Brillouin spectrum; v_{Bs} is the Brillouin frequency of the resonance, and v_B is the central frequency of the Brillouin spectrum. The dip amplitude is defined as the Rayleigh Equivalent Criterion (REC).

Figure 8 represents experimental data obtained in controlled laboratory conditions. A section of 1.5 m out of 40 m is subjected to traction by applying linearily rising weight. The unstrained fibre Brillouin frequency is 12819.98 MHz. By introducing the REC, the minimum spatially resolvable stress section is reduced to ½ the spatial resolution with the Brillouin frequency uncertainty of 5% compared to the normal spatial reolsution, which is pulse width. Apparently, the REC is an efficient threshold to unambiguously detect stress sections that are shorter than spatial resolution with an uncertainty lower than 5%. Apparently using REC through the computer program without introducing hardware change, the location accurcay can be improved by a factor of two, yet it avoids the broadband electronics for the detectors and amplifers.

7. Applications of the Distributed Sensors for Civil Structural Applications

To detect faults and assess the severity of the damage for the maintenance and repair of civil structures at the appropriate time requires monitoring the structural and local strains, which allows determination of the status of the structures. The distributed Brillouin sensor technique has the advantage of providing the information required by civil applications. Recent studies (Zeng et al., 2002, Murayama et al., 2003) have been conducted to monitor structural strain of composite and concrete beams under limited load, i.e. the structure responds to the load linearly with the average strain over the spatial resolution being monitored. The extraction of average strain used single peak fitting or centroid method (integrate the spectrum area method) (Horiguchi et al., 1995, Zeng et al., 2002, Murayama et al., 2003, DeMerchant et al., 1999, Bao et al., 2001). Those methods provide fast signal processing to identify the large strain points, while multiple peak detection (Zou et al., 2006) provides better strain accuracy with a compromise of the long processing time which makes it difficult for dynamic detection and disasters prevention. If the dynamic strain or temperature monitoring is required, the

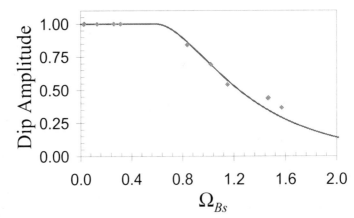

Figure 8. Normalised Brillouin loss spectrum dip as a function of the normalised Brillouin frequency shift. Simulation results (plain curve) and experimental data (diamond) at the middle of sensing length 40 m, Pp = 5 mW, Pcw = 3 mW, pulse length of 20 cm, ER = 11 dB. From Fabien Ravet PhD Thesis (2007).

intensity detection at fixed Brillouin scattering could be used without scanning the Brillouin spectrum (Bao et al., 1996), which saves the measurement time significantly. The first four civil structural applications are under the elastic working region with the purpose of verifying the distributed Brillouin sensor for strain monitoring, in which no deformation happened during the loading process, while the last section was designed using distributed sensor to monitor the deformation process, so that the strain distribution could be used to understand the deformation formation process and special feature of the deformation along the structures. Most of the recent research of the distributed sensor has been focused on using distributed Brillouin sensor as a tool to correlate the strains to the civil structural condition, such as cracks and deformation (Ravet and Bao, 2006a, Ravet et al., 2006b), as well as prediction of the crack of concretes and buckling in pipelines using special signal processing schemes.

7.1. DYNAMIC STRAIN OR TEMPERATURE MONITORING

For the distributed Brillouin sensor system based on the Brillouin gain or loss mechanism or BOTDR, in order to get the Brillouin spectrum the frequency scanning in the optical or electrical domain (coherent Brillouin scattering) is required. This procedure slows down the data processing. To reduce the measurement time, one approach is to fix the frequency difference between the pump and probe lasers at a threshold vlaue for a specfic temperature and strain, so that the relative Stokes power variation can be measured against the temperature or strain changes. This reduces the measurement time from minutes to under seconds (Bao et al., 1996) for the spatial resolution of meters over the sensing length >20 km. With intro-duction microprocessers, the sumeter spatial resolution can be achieved with 1s or less data processing time (Ferrier et al., 2004). In order to improve the power depenent temperature or strain measurement accuracy, the calibration of Stokes power or Brillouin gain coefficeint must include the temperature or strain induced bandwidth change.

Using this fixed Brillouin frequency method the "hot spot" or "high strain point" can be identified within seconds. This method can be used for threshold monitoring with lower temperature and srain resolution compared to the full Brillouin spectrum scan method.

7.2. STRAIN MONITORING OF SIMPLE CANTILEVER BEAM

Although the distributed Brillouin sensor was developed in the early 1990's, most of the measurements were demonstrated on uniform tempera-ture or strain over the spatial resolution, which is far from the real structural monitoring requirement where non-unfirom strain montioring is required. In the late 1990s the first sub-meter resolution structural strain monitoring was demonstrated on a simple cantiliver beam (DeMerchant et al., 1999) as shown in Figure 9.

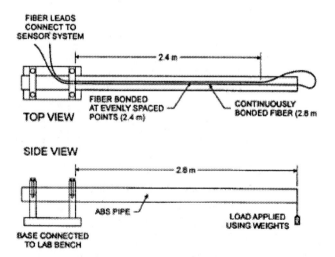

Figure 9. Top and side views of the experimental apparatus used to test the sensor system. From DeMerchant et al. (1999). ©OSA

The fiber was continuoulsy bonded to the plastic beam, and strain was recovered using the centroid method to find the average strain over the spatial resolution of 40 cm. To verify the sensor reading, electrical resistance strain gauges were also bonded to the beam at the midpoint of the beam and 0.2 m from the fixed end of the beam. The strain readings obtained from the strain gauge are well matched with the distributed sensor measurement. Figure 10 shows the distributed Brillouin sensor measured strains at different loads which were compared with the calculated strain. Both expected strain and measured strain reading are within strain accuracy of less than 50με.

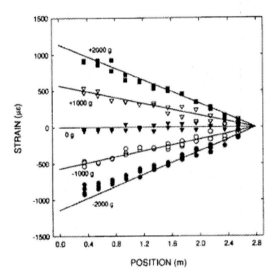

Figure 10. Strain distribution along the length of the beam at various loads as measured at 12 locations along a continuously bonded fiber by use of a 0.4-m spatial resolution. The predicted distributions are also shown with solid curves. From DeMerchant et al. (1999). ©OSA

7.3. COMPRESSIVE AND TENSION STRAIN ON THE STEEL BEAM

Distributed sensor based on Brillouin scattering could be used for tension and comrpession measurement in which the fiber is pre-tensioned, so the compression was measured in the form of released tension. However, in the civil structural applications the compression range is often unknown it would be difficult during fiber installation to apply a specific amount of pre-tension. It is important to test the standard fiber for its ability to measure the compressive strains without pre-tension applied during the installation. Furthermore, the distributed sensor measures the average strain over the spatial resolution and such a method works well with a small strain slope, i.e. low strain gradient. If the civil structure is subjected to sharp strains where the strain gradient is large and the strain slope changes along the structural locations, the response of the distributed Brillouin sensor should be checked as well as the repeatability of the sensor system.

To answer the above questions and concerns, a steel beam of 7.3-m long, 25 × 25 × 3.2 mm hollow structural section was designed and is subjected to a two-span loading as shown in Figure 11. The beam was 350-MPa grade steel and the maximum stress applied was under 100 MPa. The sensing fiber was polyvinyl chloride buffered single-mode fiber 28, attached to the top of the beam using epoxy-based structural adhesive. The fiber was bonded to the steel beam with minimum pre-tension, so that it would be possible to load the beam and to put the fiber under the condition of compression.

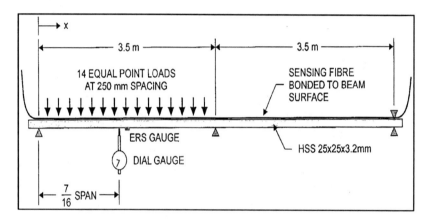

Figure 11. The steel beam under a strain distribution. From Bao et al. (2001). ©IEEE

The complete strain distribution over the entire 7-m beam at two different loads is shown in Figure 12. The strain was measured every 5 cm along the beam for a total of 141 measurements. Each data point represents the average strain over 50-cm spatial resolution along the beam, rather than the strain at one particular point. This explains the difference between the experimental (average over the pusle length) and the calculated (local) strains at the sharp turn in the tension sections for both loads.

A least square line was fitted to the load-strain data obtained at each of the 141 measurement positions along the beam, and the residuals between each data point and the line were calculated. The standard deviation of the residuals was 10με.

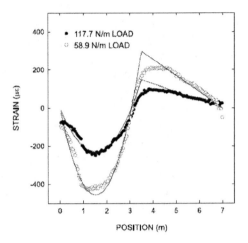

Figure 12. Measured strain distributions along the beam of two different load levels. From Bao et al. (2001). ©IEEE

On the Brillouin-scattering spectrum the compression and tension peaks have similar shapes, which means it is possible to apply the fiber in loose state to measure the compression. Hence, the tensile-strain measurement range of the distributed sensor system would be increased because no pre-tension would be needed during fiber installation. In addition to this advantage, some other applications such as pressure in the oil well can benefit from this result where large dynamic strain is required (Zou et al., 2006), for SMF28, the strain range is 15000με (including compression and tension).

7.4. CONCRETE BEAM WITH FIBERS EMBEDDED IN GLASS FIBER REINFORCED POLYMER RODS TO PROTECT BARE FIBERS

In civil structural monitoring, the optical fibers are glued to the surface of the various structures. If the fiber is embedded inside the concrete beam, then the bare optical fiber is quite fragile in a reinforced concrete structure. It is difficult to keep the optical fiber strand straight inside the concrete beam. Any bend or kink of the fiber induced by the concrete pouring process would result in large signal loss and reduces the sensing capability of the fiber. Therefore it is important to protect the fiber for distributed strain monitoring along the structures. In addtion, the bare optical fiber is also prone to chemical damage owing to the high alkali content of the concrete. The chemical reactions may occur between the bare fiber and the concrete materials which affects the sensing capability. Here the optical fiber was embedded in a glass fiber reinforced polymer (GFRP) rod acting as reinforcing bar in concrete beam, and then the optical fibers senses the internal strain of the concrete structure. There are two rods immbeded in the concrete beam: one on the top beam for comrpession measurement, labeled as CR, and one on the bottom beam for the tension measurement, labeled as TR).

The reinforced concrete beam was 1.65 m long with dimensions 100 mm × 200 mm, and it incorporateed two methods of optical fiber protection. The first method involved the embedding of fibers within pultruded GFRP rods to increase their rigidity and resistance to mechanical and chemical damage. The second method consisted of bonding the fibers to steel reinforcing bars with epoxy adhesives to prevent the fibers from reacting with concrete materials of high alkalinity. Figure 13 shows the layout of the rebars and rods inside the formwork of the concrete beam.

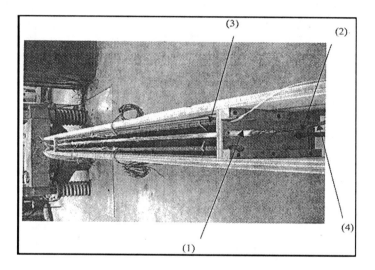

Figure 13. Layout of the rebars and rods inside the formwork of the concrete beam. 1: Top GFRP rod, CR, contains the optical fiber in the middle of the rod. 2: Bottom GFRP rod, TR, contains the optical fiber in the middle of the rod. 3: Top rebar, CS, the testing device on which the optical fiber is bonded. 4. Bottom steel reinforcing bar, TS, the testing device on which the optical fiber is bonded. From Zeng et al., 2002. ©OSA

Figures 14 and 15 show the compression and tension of the distributed Brillouin sensor through GFRP rods and fiber bonded on the steel bars and their comparison with the strain gauge and mechanical gauge. The strain from the fiber embedded in the GFRP gave the lower strain as it was protected by the GFRP bar. Both methods, fiber embedded in GFRP rods and fiber bonded to steel reinforcing bars, were found to effectively protect the optical fiber strand. The distributed fiber sensor strain resolution is less than 20 cm for the spatial resolution of 50 cm.

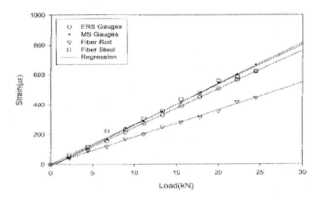

Figure 14. Comparison of one-point load tensive strains acquired from different strain gauges. From Zeng et al., 2002. ©OSA

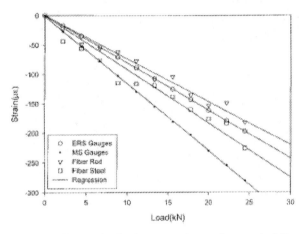

Figure 15. Comparison of two-point load compressive strains acquired from different strain gauges. From Zeng et al., 2002. ©OSA

7.5. PIPELINE BUCKLING MONITORING

The demand of oil and gas requires large pipe and higher operating pressure for higher flow capacity of pipeline. Buried land and offshore pipelines commonly suffer upheaval buckling where the pipe is extricated vertically from the protective trench, especially in northern regions where permafrost is encountered. It ultimately generates buckling by the combined effect of large temperature and pressure changes. Therefore, pipeline integrity must be monitored to prevent any leakage that can lead to costly situations, such as a disruption of the energy supply chair, environmental degradation and tarnishing of industry's image.

The tested specimen was a steel pipe of 1 m with a diameter of 18 cm and square end caps (20 cm side length). The pipe wall was thinned at the mid-length to induce weakness in the structure for buckling to occur under an axial load (Ravet et al., 2006).

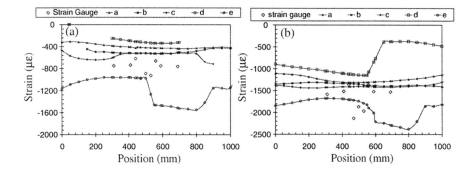

Figure 16. Pipeline Strain profiles obtained by Brillouin sensor and strain gauges measurement for (a) the 350 kN load applied and (b) the 700 kN load applied. Labels a, b, c, d and e report the measurements obtained with the Brillouin sensor on fibers a, b, c, d and e. From Ravet et al., 2006. ©IEEE

Figure 16 shows the strain along the axial direction of the pipe under 350 kN and 700 kN loads. Strain gauge readings and Brillouin sensor measurements appear to lie in the same strain range for a given load. Material non-uniformity along axial directions can also be measured by collecting data along different fibers on the axial direction. To check the elastic property of the compression and tension is another effective method of identifying the starting of the deformation at an earlier state. More specifically, in the axial direction, it appears from Figure 16(b) that curves e and d represent increased and decreased compression, respectively, in the region 600 and 800 mm, while sections a, b, and c have a uniform compressive strain, the combined tension and compression is an indication of the pipe buckling (Ravet et al., 2006). This feature has been confirmed by another pipe buckling test where the pipe is made of the high strength material with large diameter. The compression and tension appeared in the buckling point simultaneously (Zou et al., 2006). When the pipe is deofrmed or being buckled, the compression or tension on the pipe is changing nonlinearly with the load as shown in Figure 17 and the Brillouin spectrum width is bordened significantly. This may serve as another indication of the buckling.

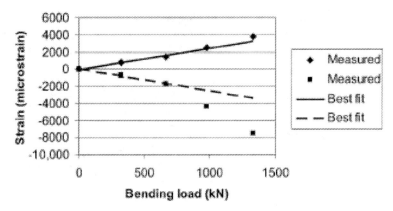

Figure 17. Strain–load relation at the buckling location. The elastic coefficients are the same for both compression and tension. However, the buckling behavior on the tensile side is different from that on the compressive side, and the localized pipe-wall buckling would happen on the compressive side prior to on the tensile side. From Zou et al., 2006. ©OSA

Through tests on a few civil structures with distributed Brillouin sensing, it is shown that the distributed Brillouin sensor allows the detection of deformations, buckling and cracks and it can provide early detection of the structural failures and can thus warn the engineer of impending failure. It strengthens the distributed Brillouin sensor as a serious practical candidate for SHM applications.

Acknowledgement

The authors would like to thank Dr. John Cameron for reading drafts and making helpful suggestions.

References

Afshar, V.S., Ferrier, G.A., Bao, X., Chen, L. (2003). Effect of the finite extinction ratio of an electro-optic modulator on the performance of distributed probe-pump Brillouin sensor systems. Opt. Lett., 28, 1418–1420.

Alahbabi, M.N., Cho, Y.T., Newson, T.P. (2006). Long-range distributed temperature and strain optical fibre sensor based on the coherent detection of spontaneous Brillouin scattering with in-line Raman amplification. Meas. Sci. Technol., 17, 1082–1090.

Bao, X., Webb, D.J., Jackson, D.A. (1993). 32-km distributed temperature sensor based on Brillouin loss in an optical fiber. Optics Letters, 18, 1561–1563.

Bao, X., Webb, D.J., Jackson, D.A. (1994). Combined distributed temperature and strain sensor based on Brillouin loss in an optical fiber. Optics Letters, 19, 141–142.

Bao, X., Webb, D.J., Jackson, D.A. (1996). Distributed temperature sensor based on "Brillouin loss in an optical fiber for transient threshold monitoring", Can J. Phys., 74, 1–3.

Bao, X., Yu, Q., Chen, L. (2004). Simultaneous strain and temperature measurements with PM fibers and their error analysis using distributed Brillouin loss system. Optics Letters, 29, 1341–1344.

Bao, X., Brown, A., DeMerchant, M., Smith, J. (1999). Characterization of the Brillouin-loss spectrum of single-mode fibers by use of very short (<10-ns) pulses. Optics Letters, 24, 510–512.

Bao, X., Yu, Q., Kalosha, V.P., Chen, L. (2006). The influence of prolonged phonon relaxation on the Brillouin loss spectrum for the nanosecond pulses. Opt. Lett. 31, April 888–890.

Bao, X., Dhliwayo, J., Heron, N., Webb, D.J., Jackson, D.A. (1995). Experimental and theoretical studies on a distributed temperature sensor based on Brillouin scattering. Journal of Lightwave Technology, 13, 1340–1348.

Bao, X., DeMerchant, M.D., Brown, A., Bremner, T.W. (2001). Tensile and Compressive Strain Measurement in the Lab and Field With the Distributed Brillouin Scattering Sensor. Journal of Lightwave Technology, 19, 1698–1704.

Boyd, R.W. (2003). Nonlinear Optics. Academic Press (Second edition). Chap. 8.

Brown, A., DeMerchant, M., Bao, X., Bremner, T.W. (1999). Spatial resolution enhancement of a Brillouin distributed sensor using a novel signal processing method. IEEE J. Lightwave Technol., 17, 1179–1183.

Chen, O., Wan, Y., Zou, L., Bao, X., Chen, L. (2004). Development of the offset locking based distributed sensor. Photonics North, Ottawa.

Culshaw, B., (2004). Optical fiber sensor technologies: opportunities and-perhaps-pitfalls. Journal of Lightwave Technology, 22, 39–50.

Culverhouse, D., Farahi, F., Pannell, C.N., Jackson, D.A. (1989). Potential of stimulated Brillouin scattering as sensing mechanism for distributed temperature sensors. Electronics Letters, 25, 913–915.

DeMerchant, M.D., Brown, A., Bao, X., and Bremner, T.W. (1999). Structural Monitoring by use of a Brillouin Distributed Sensor. Applied Optics, 38, 2755–2759.

Doi, Y., Fukushima, S., Ohno, T., Yoshino, K. (2001). Frequency stabilization of millimeter-wave subcarrier using laser heterodyne source and optical delay line. IEEE Photonics Technology Letters, Vol. 13, No. 9. 1002–1004.

Fellay, A., Thévenaz, L., Facchini, M., Niklès, M., Robert, P. (1997). Distributed sensing using stimulated Brillouin scattering: towards ultimate resolution. OSA Technical Digest Series, 16, 324–327.

Ferrier, G., Bao, X., Zou, L., Chen, L. (2004). Distributed Brillouin temperature spectra measurement without frequency scanning for dynamic process monitoring. SPIE Smart Structures/NDE Joint Conference, Nondestructive Evaluation and Health Monitoring of Aerospace Materials and Composites III, San Diego, California USA. V. 5393-10. pages: 66–75.

Horiguchi, T., Tateda, M. (1989). Optical-fiber-attenuation investigation using stimulated Brillouin scattering between a pulse and a continuous wave. Optics Letters, 14, 408–410.

Horiguchi, T., Kurashima, T., Tateda, M. (1989). Tensile strain dependence of Brillouin frequency shift in silica optical fibers. IEEE Photonics Technology Letters, 1, 107–108.

Horiguchi, T., Kurashima, T., Tateda, M., Ishihara, K., Wakui, Y. (1992). Brillouin characterization of optical fiber strain in bent slot-type optical-fiber cable. Journal of Lightwave Technology, 10, 1196–1201.

Horiguchi, T., Shimizu, K., Kurashima, T., Tateda, M., Koyamada, Y. (1995). Development of a distributed sensing technique using Brillouin scattering. Journal of Lightwave Technology, 13, 1296–1302.

Hotate, K., Hasegawa, T. (1998). Measurement of Brillouin Gain Spectrum Distribution along an Optical Fiber with a High Spatial Resolution using a Novel Correlation-Based Technique – Demonstration of 45 cm spatial resolution. OSA Technical Digest Series 16, 337–340.

Hotate, K., Hasegawa, T. (2000). Measurement of Brillouin gain spectrum distribution along an optical fiber with a high spatial resolution using a correlation-based technique – Proposal, experiment and simulation, IEICE Trans. Electron., E83 C(3), pp. 405–411.

Hotate, K., Tanaka, M. (2002). Distributed fiber Brillouin strain sensing with 1cm spatial resolution by correlation-based continuous-wave Technique, IEEE Photon. Tech. Lett., Vol. 14, No. 2, pp. 179–181.

Huai, H.K., Lees, G.P., Newson, T.P. (2000). All-fiber system for simultaneous interrogation of distributed strain and temperature sensing by spontaneous Brillouin scattering. Optics Letters, 25, 695–697.

Kalosha, V.P., Ponomarev, E., Chen, L., Bao, X. (2006). "How to obtain high spectral resolution of SBS-based distributed sensing by using nanosecond pulses," Opt. Express, 14, 2071–2078.

Kurashima, T., Horiguchi, T., Tateda, M. (1990). Distributed-temperature sensing using stimulated Brillouin scattering in optical silica fibers. Optics Letters, 15, 1038–1040.

Lecoeuche, V., Webb, D.J., Pannell, C.N., and Jackson, D.A. (1999). Transient response in high-resolution Brillouin-based distributed sensing using probe pulses shorter than the acoustic relaxation time. Opt. Lett. 25, 156–158.

Li, Y., Zhang, F., Yoshino, T. (2003). Wide temperature-range Brillouin and Rayleigh optical-time-domain reflectometry in a dispersion-shifted fibre. Applied Optics, 42, 3772–3775.

Maughan, S.M., Kee, H.H., Newson, T.P. (2001). 57-km single-ended spontaneous Brillouin-based distributed fiber temperature sensor using microwave coherent detection, Optics Letters, 26, 331–333.

Murayama, H., Kageyama, K., Naruse, H., Shimada, A., and Uzawa, K. (2003). Application of fibre-Optic distributed sensors to health monitoring for full-scale composite structures. Journal of Intelligent Material Systems and Structures, 14, 3–13.

Ohno, H., Naruse, H., Kihara, M., Shimada, A. (2001). Industrial Applications of the BOTDR Optical Fiber Strain Sensor. Optical Fiber Technology, 7, 45–64.

Parker, T.R., Farhadiroushan, M., Feced, R., Habderek, V.A. (1998). Simultaneous distributed measurement of strain and temperature from noise-initiated Brillouin scattering in optical fibers. IEEE Journal of Quantum Electronics, 34, 645–659.

Ravet, F., Bao, X. (2006a). Signatures of structure failure using asymmetric and broadening factors of Brillouin spectrum, IEEE Photonics Technology Lett. 18, January-February 394–396.

Ravet, F., Bao, X., Yu, Q., Chen, L. (2005). Criterion for sub-pulse-length resolution and minimum frequency shift in distributed Brillouin sensors. IEEE Photon. Techno. Lett. 17, 1504–1506.

Ravet, F., Zou, L., Bao, X., Chen, L., Huang, R.F., Khoo, H.A. (2006b). Detection of buckling in steel pipeline and column by the distributed Brillouin sensor, Optic Fiber Technology, V. 12, 305–311.

Ravet, F. (2007). Brillouin spectrum properties and their implications on the distributed Brillouin sensor. Ph.D Thesis.

Snoddy, J., Li, Y., Ravet, F., Bao, X. (2006). Stabilization of EOM bias voltage drift using lock-in amplifier and PID controller in distributed Brillouin sensor system. Applied Optics. In press.

Tateda, M., Horiguchi, T., Kurashima, T., Ishihara, K. (1990). First measurement of strain distribution along field-installed optical fibers using Brillouin spectroscopy. Journal of Lightwave Technology, 8, 1269–1273.

Tennyson, R.C., Coroy, T., Duck, G., Manuelpillai, G., Mulhivill, P., Cooper, D. J.F., Smith, P.W.E., Mufti, A.A., Jalali, J.J. (2000). Fibre optic sensors in civil engineering structures. Canadian Journal of Civil Engineering, 27, 880–889.

Thevenaz, L., Pellaux, J.P., Von der Weid, J.P. (1988). All-fiber interferometer for chromatic dispersion measurements. Journal of Lightwave Technology, 6, 1–7.

Thévenaz, L., Niklès, M., Fellay, A., Facchini, M., Robert, P. (1998). Truly distributed strain and temperature sensing using embedded optical fibers. Proc. SPIE 3330, 301–314.

Yu, Q., Bao, X., Chen, L. (2004). Temperature dependence of Brillouin frequency, power and bandwidth in Panda, Bow tie and Tiger PM fibers. Opt. Lett. 29, 17–19.

Yu, Q., Bao, X., Ravet, F., Chen, L. (2005). A simple method to identify the spatial resolution better than pulse length with high strain accuracy. Opt. Lett. 30, No. 17, 2215–2217.

Zeng, X., Bao, X., Chhoa, C.Y., Bremner, T.W., Brown, A.W., DeMerchant, M.D., Ferrier, G., Kalamkarov A.L., Georgiades, A.V. (2002). Strain measurement in a concrete beam by use of the Brillouin-scattering-based distributed fibre sensor with single-mode fibers embedded in glass fibre reinforced polymer rods and bonded to steel reinforcing bars. Applied Optics, 41, 5105–5114.

Zou, L., Bao, X., Afshar, S., Chen, L. (2004). Dependence of the Brillouin frequency shift on strain and temperature in a photonic crystal fibre. Optics Letters, 29, 1485–1487.

Zou, L., Bao, X., Chen, L. (2003). Study of the Brillouin scattering spectrum in photonic crystal fiber with Ge-doped core. Opt. Lett., 28, 2022–2024.

Zou, L., Bao, X., Ravet, F., Chen, L. (2006). Distributed Brillouin optical fiber sensor for detecting pipeline buckling in an energy pipe under internal pressure. Applied Optics, 45, No. 14, 3372–3377.

AN INTRODUCTION TO RADIATION EFFECTS ON OPTICAL COMPONENTS AND FIBER OPTIC SENSORS

FRANCIS BERGHMANS[*]
SCK·CEN, Boeretang 200, B-2400 Mol, Belgium and Vrije Universiteit Brussel, Pleinlaan 2, B-1050 Brussels, Belgium

BENOÎT BRICHARD, ALBERTO FERNANDEZ FERNANDEZ, ANDREI GUSAROV, MARCO VAN UFFELEN
SCK·CEN, Boeretang 200, B-2400 Mol, Belgium

SYLVAIN GIRARD
Commissariat à l'Energie Atomique, CEA DIF, Bruyères-le-Châtel, F-91680, France

Abstract. We review the effects of ionizing radiation on various types of optical components including optical fiber sensors and summarize some of their applications in particular environments where the presence of energetic radiation is a concern.

Keywords: Radiation effects; optical components; optical fiber sensor; light-emitting diode; laser diode; vertical-cavity surface-emitting laser; Bragg grating; temperature sensor

1. Introduction

Research on nuclear radiation effects on optical components has been initiated a long time ago, in the early days essentially for military purposes. Today however advanced photonic technologies for optical communication and sensing are being introduced in many other environments where the presence of highly energetic radiation is a concern. These new application

---

[*] To whom correspondence should be addressed. Francis Berghmans, SCK·CEN, Advanced Reactor Instrumentation, Boeretang 200, B-2400 Mol, Belgium, e-mail: fberghma@sckcen.be

W.J. Bock et al. (eds.), Optical Waveguide Sensing and Imaging, 127–165.
© 2008 *Springer.*

fields include space, civil nuclear industry, high energy physics experiments and future thermonuclear fusion plasma reactors.

It is well known that any electronic or photonic component may suffer from exposure to nuclear radiation.[1] The resulting system malfunctions might have dramatic consequences on safety and/or cost. Repairing components or systems in radiation environments is indeed not straightforward as human intervention might simply not be possible due to the surrounding radioactivity. Once a satellite has been put into orbit it is also almost impossible to repair onboard equipment. One therefore has to ensure that components and systems will survive the entire mission and hence one needs to carefully investigate how radiation affects their operation and their reliability.

To introduce the subject we first recall the underlying physics of radiation-matter interactions in Section 2. This section should help the reader to understand that radiation goes along with ionization and displacement processes that can significantly alter material and device characteristics. The types and intensity of radiation that the device will encounter vary considerably with the application field. Section 3 therefore summarizes the main types of radiation environments in which optical components are typically used, together with a number of applications. Sections 4 to 7 then deal with radiation effects on a number of selected optical components, including optical fibers, optical emitters, optical detectors and Bragg gratings as well as other optical fiber sensors.

2. Basics of Radiation-Matter Interactions

2.1. RADIATION-MATTER INTERACTIONS

Highly energetic radiation or nuclear radiation can come in many different forms. For the application environments considered in this chapter and detailed in section 3 one usually distinguishes between purely ionizing radiation such as X- and γ-rays and particle radiation such as protons, neutrons and heavy ions. Purely ionizing radiation deposits energy in a material mainly through the creation of secondary electrons (and positrons). Particle radiation on the other hand interacts with a material both through ionization and by non-ionizing energy loss, the latter being due for instance to the displacement or the vibration of an atom. Figure 1 attempts to summarize the different particle types and their interactions with materials.

The energy of the γ-photons usually considered in radiation testing is around 1 MeV. Therefore and considering the nature of the electronic of optical materials involved (typically low atomic numbers) the main interaction for these photons occurs through Compton scattering (Figure 2). The incoming photon ejects an electron from an atom and a photon of lower energy is scattered from the atom.

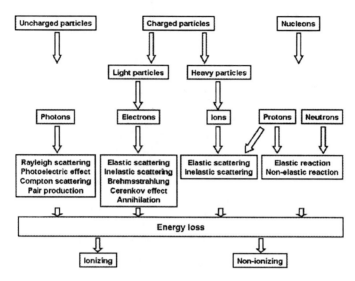

Figure 1. Schematic summary of particle types and of their interaction with materials. (Adapted from Wrobel[2]).

Nucleons such as protons and neutrons interact through elastic and non-elastic reactions. In both cases a secondary ion is produced which can in its turn ionize the material. Since protons are charged they can also yield immediate ionization through elastic or inelastic scattering processes.

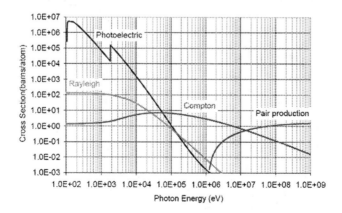

Figure 2. Partial cross sections for the different photon-material interaction mechanisms in silicon as a function of photon energy.[2]

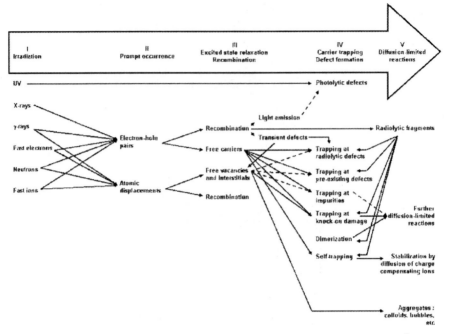

Figure 3. Cascade of radiation effects in glass material. (Adapted from Griscom[3]).

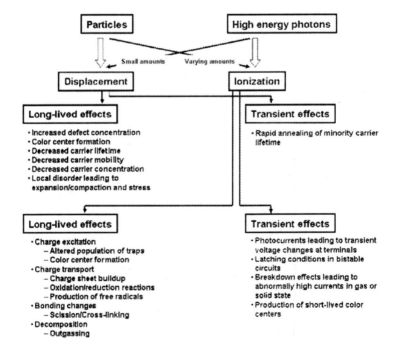

Figure 4. Schematic summary of main radiation effects applied to semiconductors. (Adapted from Holmes-Siedle[1]).

From the above it is evident that whenever a material, e.g. a dielectric or a semiconductor, is exposed to energetic radiation it can experience a cascade of effects (see for example Figure 3) which eventually lead to a modification of the electronic and atomic structure. This in its turn modifies the material properties and therefore also the characteristics of the devices manufactured from these materials, as illustrated in Figure 4 for semi-conductors.

2.2. QUANTIFYING RADIATION LEVELS AND RADIATION DAMAGE

A complete description of radiation effects requires quantifying the amount of radiation received by the components. One needs to make a distinction between ionizing energy loss, which is quantified by the concept of ionizing "dose", and Non-Ionizing Energy Loss (*NIEL*) which describes bulk damage due to atomic displacements.

2.2.1. *Total Ionizing Dose*

As seen above and generally speaking the interaction of uncharged particles (e.g. γ-rays and neutrons) with matter proceeds in two stages. First the incident particle transfers kinetic energy to charged particles in the material, and second that energy is deposited in the material as those charged particles slow down. The "Kinetic Energy Released in Matter" (*kerma*) and the ionizing "Dose" (*D*) describe these two processes, respectively. Considering Figure 5, we can define *kerma* as

$$kerma = \frac{dE_T}{dm} \qquad (1)$$

where E_T is the sum of the initial kinetic energies of all charged ionizing particles created by uncharged ionizing particles in the material. The unit of *kerma* is J/kg or Gy (Gray) where 1 Gy ≡ 1 J/kg. This quantity must be defined with respect to the specific material in which the interaction occurs, e.g. air *kerma*, water *kerma*, etc. In the definition of *kerma* the secondary particle does not need to stop inside the material volume.

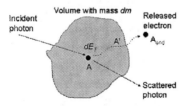

Figure 5. kerma for a single photon interacting through Compton scattering and transferring an energy dE_T to an electron released in A. The electron deposits energy along its track A' and slows down to stop in A_{end}.

Whereas *kerma* stands for the energy released per unit mass in a material, D stands for the energy absorbed per unit mass in that material. One then defines

$$D = \frac{dE}{dm} \tag{2}$$

where dE is the mean energy imparted by ionizing radiation to matter of mass dm. The total ionizing dose D is also expressed in Gray. The energy imparted by ionizing radiation to the matter in a volume is the difference between the incident radiant energy, i.e. the sum of the energies (excluding rest energies) of all charged and uncharged ionizing particles which enter the volume and the emerging radiant energy, i.e. the sum of energies (excluding rest energies) of all those charged and uncharged particles which leave the volume; the sum of all changes of the rest mass energy of nuclei and elementary particles in any nuclear transformation which occur in the volume is to be added. When charged particle equilibrium exists in the material at the point of interest, and when brehmsstrahlung loss is negligible as well as the kinetic energy of the particles produced by nuclear reactions, the *kerma* equals the dose at that point.

$$D = kerma - B \cong kerma = \int_{0}^{\infty} KF_E \cdot \phi_E \cdot dE \tag{3}$$

where B is the brehmsstrahlung energy, KF_E is the energy dependent *kerma*-factor and ϕ_E is the particle fluence.

2.2.2. *Non-Ionizing Energy Loss*

The degradation of semiconductor devices caused by radiation damage to the bulk is roughly proportional to the amount of displacement damage (creation of vacancies & interstitials), measured in terms of the total kinetic energy imparted to displaced semiconductor atoms. This bulk damage is usually expressed in terms of the damage that would be caused by a given flux of neutrons. Since the kinetic energy imparted to recoil nuclei by incident neutrons depends strongly on the neutron energy, it is conventional to express bulk damage relative to the damage that would be caused by a given flux of 1 MeV neutrons. This is for example formalized in the American Society for Testing and Materials (ASTM) standard E772.

The observation that bulk damage is proportional to the total kinetic energy imparted to displaced atoms is called "the *NIEL* hypothesis". More generally, any particle fluence can be reduced to an equivalent 1 MeV neutron fluence producing the same bulk damage (i.e. the same amount of displacements) in a specific semiconductor. *NIEL* is expressed in units of keV·cm^2/g. The *kerma* is then calculated by multiplying the *NIEL* value by the incident fluence and by the weight in grams of the irradiated material sample, as in Eq. (4).

$$kerma[keV] = NIEL\left[keV \cdot \frac{cm^2}{g}\right] \times \phi\left[\frac{1}{cm^2}\right] \times weight[g] \qquad (4)$$

The same quantity is often given in terms of the "displacement damage cross section" D_d, expressed in units of MeV·mb. To compute the total *kerma* from D_d, one multiplies by the incident fluence and by the number of irradiated atoms N, remembering the definition of a barn (or "millibarn" mb):

$$kerma[keV] = D_d[MeV \cdot mb] \times \phi\left[\frac{1}{cm^2}\right] \times N \times 10^{-27}\left[\frac{cm^2}{mb}\right] \qquad (5)$$

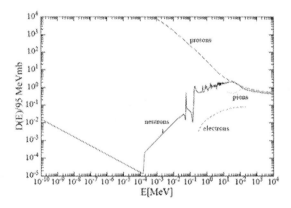

Figure 6. Displacement damage in silicon for neutrons, protons, pions and electrons.[5]

The displacement damage caused by neutrons of energy close to 1 MeV is a very strong function of energy (see for example Figure 6 for silicon). A value of 95 MeV·mb is commonly used to calculate the equivalent 1 MeV neutron fluence for an irradiation as specified by ASTM. The damage efficiency of any particle with a given kinetic energy E can then be described by a hardness factor k with

$$k = \frac{D_{d,particle}}{D_{d,1MeVneutrons}} \qquad (6)$$

3. Radiation Environments

3.1. SPACE

The main sources of energetic particles that are of concern to spacecraft designers are (1) protons and electrons trapped in the Van Allen belts (see for example Figure 7), (2) heavy ions trapped in the magnetosphere, (3) cosmic ray protons and heavy ions, and (4) protons and heavy ions from solar flares. The levels of all of these sources are affected by the activity of the sun. The energies range from keV to GeV and beyond.[7]

Cumulative long term ionizing damage due to protons and electrons can cause devices to suffer threshold shifts, increased device leakage (and power consumption), timing changes, decreased efficiency and functionality, optical loss, etc. Device shielding can help and electrons can be effectively attenuated by aluminum shielding even at high energies. However, while aluminum shielding is effective for low-energy protons, it is ineffective for high-energy protons (> 30 MeV).[8]

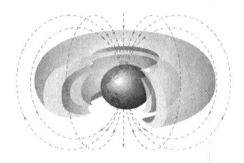

Figure 7. Artist view of the radiation belts formed by clouds and currents of particles that are trapped in Earth's magnetic field. (From NASA's "Radiation belt modeling for living with a star"[6]).

Optical components are used in spacecrafts for different purposes. A popular space application of LEDs and photodiodes is to be found in optocouplers.[9] Intra-satellite free-space or wireless optical communications are considered as well.[10] Fiber-optic gyroscopes are also used in spacecraft navigation.[11,12] Fiber Bragg gratings have been considered for onboard sensing, for example in nuclear and smart space propulsion systems,[13,14] as well as for space fiber optic communication systems[15]. Different types of optical fibers have also been evaluated for use in space based optical communications and laser ranging systems.[16,17] Many other application examples have been published throughout the years in a series of SPIE Proceedings on Photonics for Space Environments.[18–30]

3.2. HIGH ENERGY PHYSICS EXPERIMENTS

Optical components have also found their way in data communication systems for high energy physics experiments as well as for radiation dose and beam loss measurements in the vicinity of accelerators.[31] Optical links, with their favorable characteristics of high bandwidth, low power, and noise immunity, will be used extensively in the new generation of High Energy Physics experiments, such as the Compact Muon Solenoid[32] (CMS) and ATLAS[33], that are under construction at CERN for operation at the Large Hadron Collider (LHC). Optical displacement sensors have also been studied for use on the LHC superconducting dipoles.[34]

CMS is a spectacular example. Three fiber-optic systems are being implemented.[35,36] The first is an analogue optical link system of over 40000 fiber channels for readout of the CMS Tracker, where each optical fiber carries time-multiplexed data from 256 silicon micro-strips. The second system is for readout of data from the CMS Electromagnetic Calorimeter and Preshower detectors, using over 10000 fiber channels and 1 Gbit/s digital data transmission. The third system is a bi-directional 80 Mbit/s digital optical link system, with over 7000 fiber channels, for timing and control of several CMS sub-systems. The operating environment in CMS is unprecedented in High Energy Physics experiments, as a consequence of the LHC having a much higher energy, beam intensity and collision rate compared to existing accelerator facilities. Ionizing radiation doses of up to 100 kGy and particle fluences up to $2 \cdot 10^{14}$ cm^{-2}, dominated by ~200 MeV pions, are expected over the first 10 years of operation.[36] In view of the proposed upgrade of the LHC machine towards Super-LHC, as well as upgrades of the CMS detector, the effect of a 10-fold increase in luminosity is an increase to dose levels of 1 MGy and fluences in excess of $2 \cdot 10^{15}$ cm^{-2}. More information on radiation hardness issues for future high energy physics experiments have been detailed by Gill.[37]

3.3. THERMONUCLEAR FUSION PLASMA ENVIRONMENTS

We can distinguish two types of applications of optical components in thermonuclear fusion plasma environments. The first is optical diagnostics. Fiber optic spectroscopy and infrared thermography have indeed been proposed for the future International Thermonuclear Experimental Reactor (ITER)[38]. Fiber optic plasma and laser diagnostics are also considered for the Laser Mégajoule (LMJ)[39] a second application is data-communication. Fiber-optic links have been considered for communication between remote-handled tools and the control room during maintenance operations of the

ITER reactor, as well as for distribution of timing signals or triggering in LMJ experiments.

In terms of radiation environments for ITER we first need to make a distinction between the operation and the shut-down periods of the reactor. ITER's design is based on the "Tokamak" principle whereby a plasma at 100 million degrees is confined within a strong magnetic field. ITER will produce 500 MW of fusion power (10 times the power required to maintain plasma temperature). During plasma burn and due to the deuterium-tritium (DT) reaction in the reactor typical neutron fluxes in relevant locations vary between 10^{13} to 10^{17} m^{-2}s^{-1}, of which a significant fraction is 14 MeV neutrons. A fluence of between 10^{20} to 10^{24} m^{-2} over the lifetime of ITER is expected. The gamma-dose rate varies between 10^{-2} to 10^{2} Gy/s in the relevant locations during plasma operation, and up to 1 Gy/s continuously. During shut-down and maintenance periods there are no neutrons to be taken into consideration, but due to the activation of the surrounding materials and immediately after plasma shutdown dose rates exceeding 1 kGy/h are to be expected. Typical dose rates in the divertor region reach 100 to 200 Gy/h. Increased activation near the core of the plasma in the equatorial region results in expected dose rates around 500 Gy/h.[40]

The future LMJ is designed to demonstrate ignition and even gain of a DT fusion target. This facility will focus near up to 1.8 MJ UV laser light energy into a volume of less than 1 cm^3 in a few nanoseconds. A successful DT target will produce 10 times the initial laser energy by thermonuclear fusion reactions. The high-energy 14 MeV DT neutrons emitted (up to 10^{19} cm^{-2}) will interact with all the facility hardware surrounding the target creating secondary energetic gamma rays. In a few hundreds of nanoseconds the entire facility experimental hall will be filled with a mixture of nuclear radiations (DT and downscattered neutrons, gamma rays). Any instrumentation needed to operate or diagnose the target must be set inside the target chamber hall within the biological shield (30 m in diameter) and will be submitted to a moderate total dose (up to 100 Gy) but in a very short time (300 ns), i.e. at a very high dose-rate of up to 10^{10} Gy/s.[41]

3.4. CIVIL NUCLEAR INDUSTRY

Civil nuclear industry essentially encompasses the complete nuclear fuel cycle and therefore the range of possible fiber applications both for communications and sensing is very broad.[42] In 1992 already a French task force known as CORA 2000 listed valuable applications for next generation nuclear power plants. Conclusions of this group were that fiber optic sensors will present a great interest for some well-defined applications.

TABLE 1. Examples of radiation conditions that can be encountered for various applications in civil nuclear environments[1,49,50]

Application	Environment
In reactor core	Neutron flux up to 10^{14} cm^{-2}s^{-1}
	Gamma dose-rate up to 10^7 Gy/h
In containment building	Normal operation dose 5 10^5 Gy (40 years)
	Accidental dose 1.5 10^6 Gy
Fission reactor maintenance	Dose-rate 10 Gy/h, total dose 10^4 Gy
Hot-cell operations	Dose-rate 0–10^4 Gy/h, total dose 0–10^6 Gy
Decontamination work	Dose-rate 10^{-2}–10 Gy/h, total dose 10–10^3 Gy
Spent fuel manipulation	Dose-rate 10^2–10^3 Gy/h, total dose 10^6–10^7 Gy

Examples included thermal monitoring of stator copper bars in generators, structure and containment building integrity monitoring, transformers monitoring (temperature, dissolved H_2 or combustible gases in transformer oil), detection of valve positions, leak detection of primary system pressurized water or secondary system steam (e.g. in-service piping and pressure vessel systems detection of hot spots caused by small cracks) and surveillance of nuclear waste repositories.[42–45] In the meantime fiber sensors have for example been applied for temperature monitoring and distributed temperature sensing in and around experimental reactors.[46,47] Distributed radiation dose monitoring has been studied as well.[48]

Here also the radiation environment will strongly depend on the particular application. One can expect large differences in radiation fluxes and fluences whether one considers operation in a hot-cell or in the core of an operating reactor. Table 1 lists typical figures for various applications.[49,50]

3.5. SUMMARY OF RADIATION ENVIRONMENTS

From the above it is obvious that every radiation environment comes with its specific mixture of particles and flux levels. Figure 8 attempts to summarize typical values for the different environments described above in terms of ionizing dose-rate and mission total dose.

Depending on the application required equipment mission times may vary from 1000 hours to more than 10 years, resulting in total doses – expressed in Gy – that are about 4 to 5 orders of magnitude larger than the dose-rates – expressed in Gy/h. To conduct radiation testing in a reasonable time one therefore typically relies on accelerated testing up to total doses compatible with requirements set by the application, but at much higher dose-rates than those encountered during the actual service.

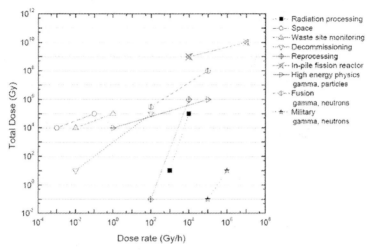

Figure 8. Summary of mission total ionizing dose vs. dose-rate for a number of application fields.

Total ionizing dose testing conventionally relies on gamma irradiation cells using ^{60}Co as radiation source emitting gamma photons with energies of 1.17 MeV and 1.33 MeV.[51] Proton or heavy ion cocktail irradiations are conducted by means of particle accelerators.[52,53] Neutron testing, depending on the required energy, can be conducted either in experimental fission reactor cores or using particle accelerators from which neutrons are generated as product of a highly energetic deuteron beam hitting a beryllium target.[52] Pulsed irradiations can be conducted using flash X-ray sources.

A number of radiation testing guidelines have been established for example by the European Space Agency[54] (ESA), by the Telecommunications Industry Association[55] (TIA) and by the International Electrotechnical Commission[56] (IEC). A military standard is available as well.[57]

4. Radiation Effects on Optical Fibers

4.1. INTRODUCTION

The effects of nuclear radiation on optical fibers and, more generally, on amorphous silica, have been extensively studied during the last three decades.[58–60] It is well-known that optical fibers can be very sensitive to ionizing radiation. Even at low total doses, the optical attenuation of the fiber can significantly increase. Radiation-induced attenuation (RIA) is primarily due to the trapping of radiolytic electrons and holes at defect sites in the fiber silica, i.e. the formation of so-called color centers.[61,62] These color centers can absorb information-carrying photons at certain wavelengths and can be made to disappear through thermal or optical processes, causing a decrease of RIA. Generally speaking, the induced attenuation is lower at higher optical wavelengths. Color centers indeed essentially absorb in the UV and visible part of the spectrum, with absorption tails extending to the infrared. The structure of the main color centers in amorphous SiO_2 is shown in Figure 9 together with the associated transition energies. During irradiation, both new color center formation and recovery compete. Therefore, the magnitude of the induced attenuation also depends on the dose-rate. The parameters influencing the optical fiber's response to ionizing radiation include:

- the fiber composition and fabrication method;
- the type of radiation, dose-rate, total dose;
- the temperature, fiber wrap diameter, stress;
- its thermal treatments and irradiation history;
- the optical wavelength and power;
- the time elapsed between exposure and measurement.

When dealing with RIA in optical fibers it is therefore of primordial importance to clearly state which type of fiber and glass composition are involved, which are the environmental conditions and at what wavelength the values are measured. Typical RIA levels for a wealth of different commercially available fibers can for example be consulted in a report by Ott.[63]

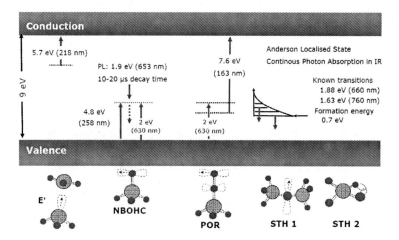

Figure 9. Illustration of main color centers in silica (NBOHC ≡ Non-Bridging Oxygen Hole Center; POR ≡ PerOxy-Radical; STH ≡ Self-Trapped Hole) and associated optical transitions.[62]

As stated earlier most fiber radiation testing happens in an accelerated manner. It is therefore very important to be able to extrapolate the high dose-rate effects to those that will be observed during service, for example for predicting the behavior of a fiber in space, for which radiation conditions cannot exactly be simulated in laboratory and where dose rates may even be time-dependent. Several RIA growth and recovery models have been proposed in literature. A model for the induced absorption growth which has been considered a couple of times in literature is a simple power-law[64]

$$RIA = \alpha D^{\beta} \tag{7}$$

where D is the total dose and α and β are empirical constants. Since recovery processes were suspected to result from bimolecular processes (e.g. diffusion-limited recombination of electrons and holes or vacancies and interstitials), the recovery after irradiation has been described by n-th order kinetics, following the relation:[65]

$$RIA = (RIA_0 - RIA_f)(1 + ct)^{-\frac{1}{(n-1)}} + RIA_f$$
$$c \equiv \frac{1}{\tau}(2^{n-1} - 1) \tag{8}$$

where RIA_0 and RIA_f are the initial and final values respectively of the induced attenuation, τ is the half-height lifetime of the recovery and n is the material-dependent kinetic order. Fractal kinetics have been used as well to study radiation-induced color center formation in silica fibers.[66] Other authors relied on a sum of saturating exponentials:[67,68]

$$RIA = \sum_i A_i \left[1 - e^{-D/B_i} \right] \tag{9}$$

where each term is attributed to a particular color center, A_i being an amplitude and B_i a saturating dose parameter, while D is the total dose. These models have been applied with varying levels of success depending on the fiber type, measurement wavelength and irradiation conditions. Recently LuValle[69] and Friebele[11] applied explicit kinetic modeling to RIA data of pure silica core polarization-maintaining fibers for space gyro applications. They showed that the growth of the induced loss is consistent with a power law in dose rate, and that recovery is consistent with low (1st or 2nd) order annealing with a distribution of activation energies and a single attempt frequency for each population.

Other radiation-induced effects in optical fibers include luminescence and densification/compaction leading to refractive index changes.[62,70,71]

These radiation-induced refractive index changes may for example affect the properties of fiber Bragg gratings (see section 7.1).

4.2. STEADY-STATE IRRADIATIONS ON PURE SILICA OPTICAL FIBERS

In the family of silica glasses (synthetic) pure silica can be considered as outstanding material for fabricating purse silica core (PSC) fibers for transmitting light in the 400–2000 nm wavelength window in a steady-state radiation environment. This wavelength range is for example important for optical plasma spectroscopy in thermonuclear fusion reactors.

As stated above RIA depends on many parameters, among which the irradiation temperature, the operating wavelength and the impurity content which certainly plays a prevailing role. An important example of a typical impurity introduced during the fiber fabrication process is chlorine. If the chlorine content is high enough it can degrade the optical transmission in the visible spectrum even if the optical band associated with this impurity peaks in the UV.

4.2.1. *Spectral Behavior of the RIA in Dry and Wet Pure Silica Fibers*

Figure 10 compares the typical RIA spectra in state-of-the-art radiation resistant dry and wet silica PSC fibers irradiated up to a total ionizing dose of 1 MGy with γ-rays. In the visible spectrum, we can note the formation of the absorption band (typically ~1 dB/m) around 600 nm (2 eV) in both type of fibers. The formation of this absorption band is primarily due to the creation of non-bridging oxygen hole centers (NBOHCs).[72]

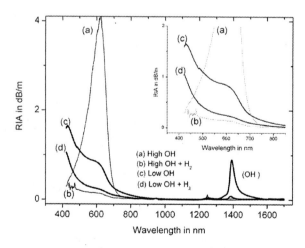

Figure 10. Typical example of RIA spectra at 1 MGy in high OH (thin line) and low OH (thick line) pure silica core fibers. The inset shows the effect of H_2 to lower the RIA at 600 nm.

The peak position of the band slightly shifts over a few tens of nanometers towards the red when the OH content decreases. This behavior was explained by the possible existence of different NBOHCs species.[73]

Above 850 nm the RIA becomes small and IR transmission can take place even at very high total dose (see section 4.2.3). The radiation resistance can be further enhanced to some extent in the visible range by relaxing the silica glass network and lowering the formation of free radicals (e.g. NBOHCs). This can be achieved by means of two methods: fluorine doping[74] or molecular hydrogen loading[75]. The former method brings outstanding radiation resistance in bulk pure silica but unfortunately core fluorine-doping reduces the numerical aperture of PSC fibers and induces additional bending loss. Therefore the hydrogen-diffusion process is usually preferred to reduce the RIA around 600 nm. Figure 10 displays an example of the efficiency of the hydrogen loading in an attempt to lower the RIA around 600 nm. In this example RIA has been reduced by a factor 4 to 10 depending on the H_2 loading content ($>10^{19}$ molec·cm^{-3}) and the fiber type.[76]

4.2.2. *Transient Absorption in Very Dry Silica Fibers*

Strong transient absorption ($>10^4$ dB/km) can also occur in the 660–760 nm region in PSC fibers irradiated in steady-state condition. Such a peculiar phenomenon is usually best observed in very dry silica prepared with a sol-gel technique or other types of glasses. This transient RIA is believed to pertain to self-trapped holes.[77,78] These defects are meta-stable and are bleached by radiation. Figure 11 and its inset show an example of such transient absorption vanishing with dose.[62]

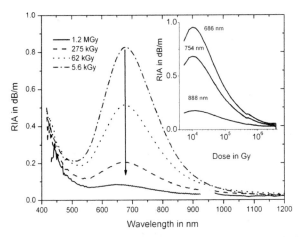

Figure 11. Spectral behavior and dose kinetics (inset) of transient RIA in low-OH/low-Cl (KS4V) F-clad pure silica core fiber (200 μm) irradiated with γ-rays at a dose-rate of ~6 Gy/s in the dark.

4.2.3. *RIA for Fission Reactor Irradiation*

In the case of reactor irradiation the intense radiation field (gamma and neutrons) causes the fiber to entirely absorb visible light, while IR transmission still remains acceptable with a minimal absorption of a few dB/m around 1000 nm even at extremely high total ionizing dose (>100 MGy).[79] Figure 12 illustrates this behavior in a dry silica. In the case of polymer-coated fibers the production of radiation-induced hydrogen, which diffuses to form OH groups creates additional absorption at 1390 nm. When a metal coating is used instead this effect is not observed.[80]

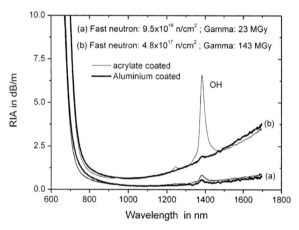

Figure 12. RIA in a dry pure silica fiber irradiated in a fission reactor. The polymer used in fiber coatings releases hydrogen which diffuses into the fiber to form OH groups absorbing light at 1390 nm.

4.3. STEADY-STATE IRRADIATIONS ON DOPED OPTICAL FIBERS

Doped core fibers are well known to be much more radiation sensitive than PSC fibers. Figure 13 gives a general comparative view of the RIA measured in similar condition in various types of doped fibers.

4.3.1. *RIA in Ge and Ge-P Co-doped Core Fibers*

Germanium is used in the fabrication process of high-bandwidth multimode or singlemode telecommunication fibers. Radiation creates Ge-associated defects, e.g GeE' and GeOHC, similar to those formed in pure silica fiber. The RIA magnitude in Ge doped fiber depends on the fabrication process as it can be shown at Figure 4. Usually, fibers fabricated by a PCVD method (characterized by a low temperature deposition process) leads to a somewhat lower RIA compared to fibers drawn from other fabrication technologies.

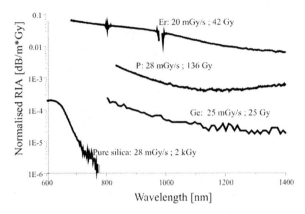

Figure 13. Comparison of normalized RIA in pure silica, Er, P and Ge-doped core fibers irradiated in similar conditions with γ–rays at 28°C.

The RIA in a fiber containing only Ge will generally follow a monotonic increase of the RIA up to saturation with substantial recovery when the irradiation is halted (Figure 14).

As long as Ge is the dominant dopant in the fibers the post-irradiation thermal annealing or recovery can be predicted from the RIA growth by applying n-th order kinetics.[81] However, when phosphorous is present as co-dopant the RIA is enhanced mainly because of a lack of recovery.[65] Figure 14 illustrates this behavior by showing the typical rebound "effect" in the recovery of irradiated Ge-P fibers (see curve 3).[62]

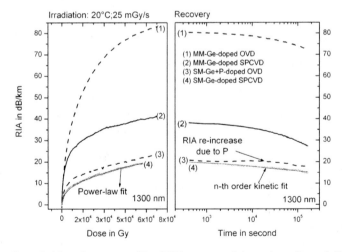

Figure 14. Growth (a) and recovery (b) of RIA measured in various Ge and Ge-P-doped fibers irradiated in similar conditions. [62]

4.4. TRANSIENT IRRADIATION EFFECTS ON OPTICAL FIBERS

Large bandwidth and electromagnetic immunity are characteristics of optical fibers that make them particularly well-suited to the requirements of fast pulsed plasma diagnostics systems.[82] The radiation-induced effects associated with these transient environments have been extensively studied during the 1980's particularly by the Los Alamos National Laboratory[83] and through international collaborations under the auspices of the Nuclear Effects Task Group reporting to NATO Panel IV.[84] More recently, researches have been motivated by the development of Megajoule class lasers and the need for radiation-tolerant waveguides for plasma and laser diagnostics (see also section 3.3).[85]

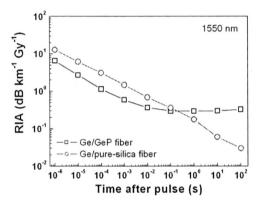

Figure 15. Comparison between the RIA for a 35 ns, X-ray (1 MeV) pulse at 1550 nm in phosphorus (P)-free and P-codoped single-mode germanosilicate fibers. RIA curves are normalized by the deposited dose (1 to 200 Gy[SiO2] range).

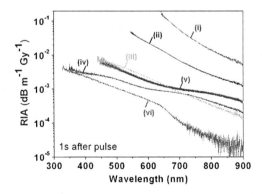

Figure 16. Radiation-Induced Attenuation (RIA) spectra measured 1 s after a 35 ns X-ray pulse (dose ~200 Gy[SiO2], dose rate>1 MGy/s), (i) P-doped core, (ii) Ge-doped core (iii) F-doped core ([F] ~100 ppm), (iv) F-doped core ([F]> 2%), (v) pure-silica-core (low-OH/low-Cl) (vi) pure-silica-core (high-OH/low-Cl). All spectra were acquired at room temperature and normalized by the deposited dose.

For transient radiation-hardening of optical links one has to carefully assess the time dependent degradation of the fiber with respect to the emission time of the radiation pulse. Figure 15 illustrates the RIA at 1550 nm measured just after a 35 ns X-ray pulse for two different types of commercial single-mode fibers. The transient RIA is very high (~10 dB/km, 1 µs after a 1 Gy pulse) compared to the ~0.2 dB/km losses pre-irradiation, providing evidence for a strong decrease of the fiber transmission.

Similarly to the case of steady-state irradiation effects, the RIA levels, the dose dependence and the decay kinetics are explained by the generation and recovery mechanisms of point defects created in the silica glass matrix. As a consequence the fiber radiation responses have been shown to be strongly dependent of parameters like core/cladding doping or drawing temperature and speed.[84,86–88] As an example Figure 16 illustrates the RIA spectra measured 1 s after an X-ray pulse for optical fibers with different core doping: pure-silica, germanium-, fluorine- or phosphorus-doped cores. The contribution of room-temperature unstable point defects explains the high transient levels measured at short times after pulse. At longer times (> 1 s), RIA is more comparable to the loss levels measured during steady-state irradiations. Previous studies have shown that unstable self-trapped excitons and holes strongly contribute to transient RIA, especially in pure-silica-core and fluorine-doped optical fibers.[89,90]

4.5. PHOTONIC CRYSTAL FIBERS

Photonic crystal fibers (PCFs) were also recently submitted to a number of radiation experiments. For these fibers one needs to distinguish solid core and hollow core versions. For solid core PCFs one can anticipate that RIA will play an equally important role as in conventional optical fibers.[91,92] In hollow core PCFs the light is essentially guided in air, which may significantly change the radiation response of such waveguides compared to conventional optical fibers. First results have shown that pulsed RIA levels in hollow core PCF are larger than those expected from the fiber structure and are quite comparable to a commercial Ge-doped single mode fiber for the dose range 0.1–10 Gy. The radiation-induced loss at 1550 nm saturated at ~0.1 dB/m for doses larger than 10 Gy.[93] The permanent RIA levels after radiation were found to be very low. This was confirmed by Henschel[94] with hollow core PCF showing at least about 30 to 100 times lower RIA than the best present conventional single mode fibers at 1550 nm.

5. Radiation Effects on Optical Emitters

Most radiation effects studies on optical emitters reported in literature rely on proton irradiations to mimic as closely as possible the space environment. It is well-known that particle radiation is substantially more damaging for emitters than gamma radiation. A major issue with emitters is the radiation-induced decrease of the minority carrier lifetime through the introduction of non-radiative recombination centers. Considering Shockley-Read-Hall theory the native lifetime τ_0 can be defined by Eq. (10):

$$\frac{1}{\tau_0} = N_d \sigma_0 v_{th} \qquad (10)$$

with N_d the defect density, σ_0 the capture cross section and v_{th} the carrier thermal velocity. At this stage the *NIEL* hypothesis can be used and one can assume the number of radiation-induced defects to be proportional to the displacement dose. The lifetime τ_{def} associated to recombination on radiation-induced defects can therefore be written as

$$\frac{1}{\tau_{def}} = kNIEL\phi\sigma_{def}v_{th} \qquad (11)$$

where k is a proportionality factor and ϕ is the particle fluence. The resulting lifetime for a given fluence $\tau(\phi)$ therefore becomes

$$\frac{1}{\tau(\phi)} = \frac{1}{\tau_0} + kNIEL\phi\sigma_{def}v_{th} = \frac{1}{\tau_0} + K\phi \qquad (12)$$

with K the "damage constant" which depends on particle type and energy.[95] For 50 MeV protons, the threshold for measurable changes in lifetime in III–V optical emitters is $10^{10}\,\mathrm{cm}^{-2}$.

Radiation damage can then affect the emitter in various ways, essentially through efficiency decrease, threshold current increase (for laser diodes), increased series resistance and accompanying thermal effects. In this section we illustrate these radiation effects as reported for light-emitting diodes (LEDs), laser diodes and vertical-cavity surface-emitting lasers (VCSELs).

5.1.1. *LEDs*

In 1982 already Rose[96] reported displacement damage results for several types of LEDs and different working conditions in a comprehensive study. The observed output power decrease can be expressed by

$$\left(\frac{L_0}{L}\right)^n = 1 + \tau_0 K \phi \qquad (13)$$

where L_0 is the initial power output, τ_0 and τ are the initial and the post-irradiation minority carrier lifetime, respectively, and ϕ is the particle fluence. n is an exponent with value between 0 and 1.

In a review paper on the subject Johnston[97] discussed that some LEDs are so sensitive to radiation damage from protons that they are severely degraded after 50 MeV equivalent proton fluences of 1 to $2 \cdot 10^{10}$ cm^{-2} (see for example Figure 17), which is a level that is experienced by many spacecraft. Amphoterically doped LEDs are very sensitive to injection-enhanced annealing. Other types of LEDs are much more resistant to radiation damage. Very high-speed devices that are intended for optical communication require radiation levels that are three orders of magnitude greater before measurable degradation occurs. A final comment is made on the care one needs to take when interpreting damage from tests involving higher proton energies as this might strongly depend on what *NIEL* value is used. The dependence of damage on particle type and energy is nevertheless critically important for space applications because the space environment consists of a wide energy spectrum of particles. Most radiation tests are done with only a single energy because of the high testing cost, interpreting the damage at different energies (or with different particles) with accepted values for relative damage. If those values are incorrect, the test results may lead to large errors in estimating the effect of the real space environment.

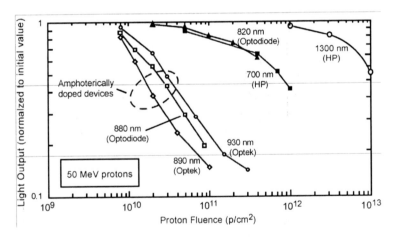

Figure 17. Decrease of the output power for several types of LEDs exposed to 50 MeV protons. (From Johnston[97]).

5.1.2. *Laser Diodes and VCSELs*

Laser diodes exposed to particle radiation are known to exhibit a shift of the threshold current. The threshold current of a laser diode can be written as

$$I_{th} = \frac{qV}{\tau_{nr}} N_{th} + qVBN_{th}^2 + qVCN_{th}^3 \qquad (14)$$

where q is the electron charge, V is the volume of the active region, B the radiative recombination coefficient, C the Auger recombination coefficient and N_{th} the carrier density at transparency, i.e. the carrier density for which the material gain is zero. The non-radiative recombination time τ_{nr} is the most likely to be affected by radiation.[98,99]

A decrease in the non-radiative recombination time thus leads to an increase in threshold current that can be expressed by

$$I_{th} = I_{th,0} + K\phi \qquad (15)$$

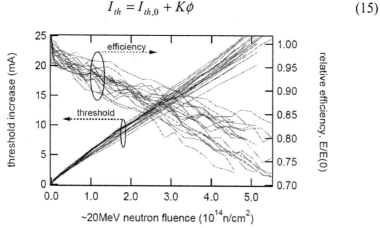

Figure 18. Example of threshold current change and normalized efficiency change for 20 identical edge-emitting laser diodes irradiated with neutrons. (From Macias[100]).

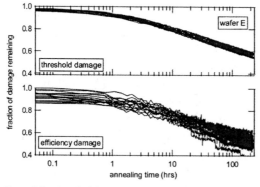

Figure 19. Annealing of the threshold current and efficiency damage after neutron irradiation for 20 identical edge-emitting lasers. (From Macias[100]).

with K a damage factor and ϕ the particle fluence. Due to the change of the carrier lifetime the emission efficiency decreases as well. This is illustrated in Figure 18.[100] Laser diodes are also known to show annealing (Figure 19[100]) which can be accelerated by the injection current.

It is remarkable that the radiation level at which the threshold currents begin to degrade is within a factor of two for different types of lasers as shown in Figure 20. The reason for this is the logarithmic dependence of the gain-current density relationship for lasers that is applicable to all three material systems used for these lasers and is essentially independent of the type of recombination process that limits cavity gain.[97]

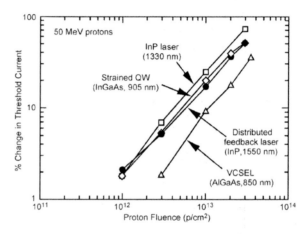

Figure 20. Threshold current degradation of several different types of laser diodes after irradiation with 50 MeV protons. (From Johnston[97]).

VCSELS form a more recent class of laser diodes. They rely on complex Bragg reflectors formed by multiple thin layers at the back region of a vertical structure. VCSELs have a very small active layer and a low threshold current but usually come with lower optical output power than conventional edge-emitting lasers. The very compact structure and the accompanying self-heating makes the operation of the structure less dependent on temperature. However, the thermal coupling causes the optical power output to drop at high currents. This limits the range of forward currents, which also means that VCSELs can tolerate smaller radiation-induced changes in threshold current compared to conventional lasers with stripe geometry. They are also affected by strong injection-enhanced annealing. VCSELs have nevertheless shown excellent radiation hardness up to total ionizing dose exceeding 20 MGy and to neutron fluences up to several 10^{14} cm^{-2}. This is illustrated for example in Figure 21 which shows the output power versus forward current in a commercially available 850 nm multimode VCSEL.[101]

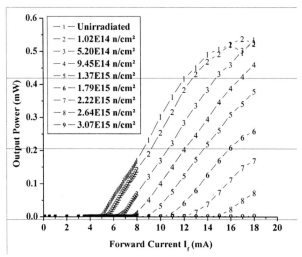

Figure 21. VCSEL L-I curves at different neutron fluence levels (expressed in neutrons per cm^2) measured at 60.0°C. [101]

Recently the total dose radiation tolerance of single mode long-wavelength VCSELs emitting in the 1.4–1.7 µm region was also investigated. These VCSELs hold promise for use in chemical and gas sensing applications. The results obtained for a 1.52 µm VCSEL are shown in Figure 22. Only limited damage was observed at a total gamma radiation dose of 10 MGy.[102]

Other radiation effects on VCSELs have been addressed by Kalavagunta[103], essentially an increase of the series resistance yielding thermal effects that in their turn cause an increase of the leakage current and a wavelength shift following 2 MeV proton irradiation. The resistivity degradation in the device was explained by carrier removal and defect-limited mobility degradation.

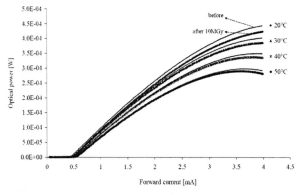

Figure 22. L-I curves measured between 20°C and 50°C before (solid lines) and after gamma irradiation up to 10 MGy (dotted lines) for a TO-46 canned VCSEL operating at 1.52 µm. [102]

6. Radiation Effects on Optical Detectors

Optical detectors and p-i-n photodiodes in particular are also affected by radiation. Displacement damage degrades minority carrier lifetime and thus diffusion length. The quantum efficiency of the detector is therefore reduced. On the other hand damage in device depletion regions increases the generation centers and the resulting charge carriers lead to increased dark current and leakage current. Displacement damage from neutron and proton irradiation in Si PDs caused the leakage current to increase by 6–7 orders of magnitude and the responsivity to decrease by 90% after 10^{15} cm^{-2}, while gamma damage was almost negligible at total doses of 10 kGy.[104] The spectral response was also reported to change under radiation.[105]

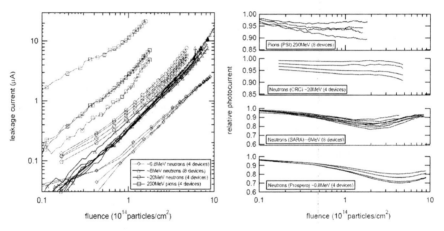

Figure 23. Damage to photodiode leakage current (left) and responsivity (right) at 5V reverse bias for identical devices at different irradiation facilities. (From Gill[106]).

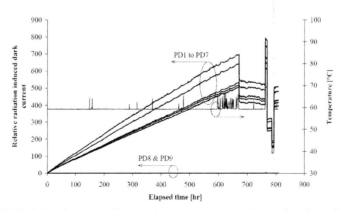

Figure 24. Relative increase of the dark currents measured as a function of time for 7 InGaAs photodiodes (PD) at a reverse bias of –5 V, a gamma dose rate ~ 15 kGy/h. PD 8 and PD 9 were used as reference, outside the irradiation container. [107]

Figure 23 shows for example the evolution of leakage current and responsivity in p-i-n photodiodes irradiated with various particle types.[106] InGaAs photodiodes were also evaluated at a total ionizing dose exceeding 10 MGy. Only little effect was observed on the responsivity, whereas the dark current substantially increased, as shown in Figure 24.[107]

7. Radiation Effects on Optical Fiber Sensors

7.1. FIBER BRAGG GRATINGS

How Fiber Bragg Gratings (FBGs) can be used as temperature, strain or pressure sensors is dealt with in detail in other chapters in this volume. The operation as temperature sensor, for example, relies on measuring the shift of the Bragg wavelength λ_B with temperature. For a restricted temperature interval this shift is well approximated by the following linear expression

$$\lambda_B(T) = \lambda_B(T_0) + \alpha_0(T - T_0) \qquad (16)$$

where α_0 is the FBG temperature sensitivity coefficient with a typical value of about 10 pm/°C at 1550 nm for Ge-doped silica fibers, and T_0 is the reference temperature. The information about temperature changes is wavelength encoded and is essentially independent on radiation-induced loss in the optical fibers, which is one of the main reasons for failure of other types of optical fiber sensors.[97]

The possible application of FBGs for temperature sensing in a radiation environment thus depends on to what extent $\lambda_B(T_0)$ and α_0 are influenced by radiation. Therefore, a radiation-induced change of the Bragg wavelength and of the grating reflectivity can not be tolerated. On the other hand, the possibility to use FBGs as radiation sensors was also discussed.[109] In that case a change of λ_B under radiation should allow to assess the radiation dose.

First attempts to assess the effect of radiation on FBGs have demonstrated that for a Ge-doped silica fiber the shift of λ_B can be as high as 0.1 nm towards the blue at a 12 kGy γ-radiation dose.[110] It was also noted that the temperature sensitivity coefficient α_0 may change under radiation. A non-monotonous shift of the Bragg wavelength under γ-radiation up to a dose of 71 kGy was also reported.[111] Those results raised doubts about the use of FBGs for sensing applications in radiation environments, since a 0.1 nm drift of the Bragg peak corresponds to a 10°C error on the temperature estimation which is unacceptable for most applications.

In subsequent publications[112–114] significantly lower values of the Bragg peak shift of ~20 pm towards longer wavelengths were reported. This redshift was in qualitative agreement with the behavior of the Bragg peak during photo-inscription of the grating and with the experimental observation that γ-irradiation of silica glass doped with 11 mol.% GeO_2 creates the same type of paramagnetic defects as those underlying the photosensitivity to 248 nm UV light.[115] However, on the quantitative level the FBGs behave differently under UV light and γ-radiation: the sensitivity to the latter is much higher, but it saturates at levels much lower than those required to write a practically meaningful grating. The difference can be explained by assuming that although UV and γ-radiation create the same types of defects, some precursors can interact with γ-rays but not with UV light. The concentration of "UV-insensitive" precursors is rather low resulting in the early saturation.

We have seen earlier that the growth of RIA in optical fibers is dose-rate dependent. The influence of the dose-rate on the Bragg peak shift of FBGs has also been studied and for gratings written in a photosensitive fiber this shift with γ-radiation was in the range from 20–80 pm for dose-rates from 1–25 kGy/h.[116] This dose-rate dependence might be responsible for some discrepancies between different published results and indicates that the physics of radiation effects in the FBG is more complicated than just simple defect generation.

FBG sensors have been demonstrated to sustain radiation loads of $2 \cdot 10^{19}$ cm^{-2} fast neutrons (>1 MeV) and a 870 MGy gamma total dose.[13] The goal of that test performed in a nuclear reactor was to assess the possible use of FBGs in future space exploration missions based on nuclear propulsion. Despite the substantial increase of the optical loss in the fiber it was possible to record grating spectra throughout the duration of the test. The Bragg wavelength shift saturated within approximately 30 minutes and the final magnitude of the wavelength shift depended upon the dose rate. The wavelength shift subsequently remained at the saturation value throughout the duration of the test.

Major components of the near-Earth space radiation environment are protons and electrons. The effect of proton radiation on the FBG was first studied by Taylor.[117] Fiber Bragg gratings written in a photosensitive fiber using a frequency-doubled Argon laser beam were exposed to 63 MeV protons with a total fluence of $7.5 \cdot 10^{13}$ cm^{-2}, which corresponds to an absorbed dose of 0.1 MGy. A Bragg peak shift due to a combined effect of the ambient temperature increase (~2.5°C) and radiation was at a level of 50 pm towards the longer wavelengths. Compensation for the temperature effect using the nominal temperature sensitivity of 8.6 pm/°C gives an estimation of the radiation-induced shift of ~20 pm. Approximately the same level of sensitivity was observed by Gusarov[112] in the case of γ-radiation with a shift saturating at a 30 kGy dose. A fast post-irradiation

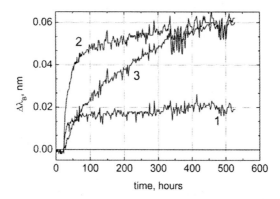

Figure 25. Shift of the Bragg wavelength $\Delta\lambda_B$ under gamma-radiation with a 3 kGy/h dose-rate for gratings written in different fibers. 1 – photosensitive fiber with 11 mol.% GeO2, written with 248 nm light, 2 – standard telecom fiber hydrogen-loaded, 330 nm light, 3 – nitrogen doped fiber, 193 nm light.

recovery was observed: 1 hour after the end of irradiation the radiation-related Bragg peak shift was within the measurement uncertainty.

The response of gratings written in a hydrogen-loaded fiber is different from that of gratings written in a naturally photosensitive fiber (Figure 25).[113] The radiation-induced shift of the Bragg peak saturated at a 0.1 MGy dose level, a higher value for FBGs written in hydrogen-loaded Ge-doped fiber than for FBGs written in unloaded Ge-doped fibers. The amplitude and the width of the Bragg peak changed during irradiation for gratings written in hydrogen-loaded Ge-doped fibers, while it did not change for gratings written in unloaded Ge-doped fibers. Changes of the grating strength during γ-irradiation were attributed to different kinetics in radiation-induced changes of refractive index at minima and maxima of the UV fringe pattern. The higher gamma-radiation sensitivity of gratings written in the hydrogen loaded fibers was thought to be due to radiolysis of OH-bonds.

In conclusion, present-day FBG sensors can be used for coarse sensing application in high radiation dose-environments. For doses below 1 kGy, which corresponds to most of near-Earth space mission scenarios, an equivalent stability better than 1°C is possible without additional precautions with gratings written in intrinsically photosensitive fibers. Development of radiation hardening techniques which would allow to use such sensors for high accuracy applications in high-dose radiation environments remains an open issue. A better insight into the mechanisms of ionizing radiation effects on the FBG would be very useful in this respect. Promising results have nevertheless shown that FBG temperature sensors still operate satisfactorily with a measured drift lower than 3°C after 50 months in a low-flux gas-cooled graphite-moderated experimental nuclear reactor.[118]

7.2. OTHER SENSOR TYPES

Various optical fiber sensor types have been looked at in terms of their radiation tolerance. Three optical fiber temperature sensors, including a semiconductor absorption, a Fabry-Perot cavity and a fluorescence type sensor were for example exposed to gamma radiation. The most gamma radiation resistant sensors appeared to be the semiconductor absorption sensors, which remained operational up to a 160 kGy dose. The Fabry-Perot sensor degraded as a function of dose, whereas the fluorescence sensors stopped operating abruptly after a 250 Gy dose. The degradation of these sensors is found to be due to their construction, more particularly to the choice of the optical fiber type. There was no evidence of any degradation of the underlying physical sensing mechanisms under gamma radiation. The semiconductor absorption sensors were also submitted to neutron radiation with a significant detrimental effect on the absorption characteristics of the semiconductor and therefore on the sensor operation.[108] Particle radiation (4 MeV α particles) has also already been reported to affect the operation of a fluorescence temperature sensor.[119]

The variation of thermo-optic coefficient caused by gamma irradiation was also investigated for a fiber Fabry–Perot interference sensor. The experiments showed that the relative change of thermo-optic coefficient increases only by 2.1% and saturates at a total dose of 300 kGy, which suggests that the fiber FFPI sensor is suitable for use in harsh gamma radiation environments.[120] Whether this remains valid at high neutron fluences remains questionable due to the anticipated compaction of the fiber silica. A recent cavity length measurement of extrinsic FFPI sensors irradiated up to a fast (E >1MeV) neutron fluence of 10^{19} cm^{-2} revealed that the absolute length of the cavity increased by 1 µm with respect to the initial value of 10 µm. This effect is mainly attributed to the neutron-induced compaction of the amorphous silica.

Raman Distributed Temperature Sensors have also been demonstrated to withstand total gamma doses in excess of 300 kGy, using a double-ended configuration and commercially-available optical fibers. No specific calibration technique is needed to correct for the radiation effects in commercially-available optical fibers, making suitable double-ended RTDS systems for temperature monitoring and fire detection in large nuclear infrastructures.[121] The effects of γ-radiation up to very high total doses on the physical properties of Brillouin scattering in standard commercially available optical fibers were also studied. A frequency variation of about 5 MHz for both Brillouin frequency and linewidth has been measured at the total dose of about 10 MGy. The radiation-induced shift has a negligible practical impact and makes Brillouin scattering very immune to radiation, so that distributed sensors based on this interaction exhibit an interesting potential for use in nuclear facilities.[122]

Whereas RIA in optical fibers is definitely a hurdle for a more widespread use of optical fibers in radiation environments, this effect has been proposed many times as the underlying operation principle for on-line fiber-optic radiation monitoring. The amount of fiber darkening can indeed be a measure for the radiation dose absorbed by the fiber. Such fiber sensors can be very small and therefore applicable to *in vivo* dose deposition measurements[123]. The principle could also be combined with optical time domain reflectometry techniques suitable for distributed dose monitoring and surveillance of large scale nuclear facilities.[31,124] Depending on the radiation level the radiation sensitivity can be adapted by adjusting the dopant concentration. For a medium dose range (10 Gy – 10 kGy) a phosphorous core doped fiber without the addition of any other co-dopant allows to minimize the post-irradiation annealing effect and therefore to maintain the dose information.[125] In the case of radiotherapy the radiation sensitivity must be increased and lead (Pb) doped fiber appears to be a better choice. PMMA fibers have also been evaluated under radiation. These allow obtaining a broad radiation sensitivity range on one single fiber by interrogating it at multiple wavelengths in the visible spectrum.[126] Other (extrinsic) measuring principles relying on luminescence in a material attached to the fiber tip have also been considered. The light emission can stem from Optically Stimulated Luminescence (OSL)[127] or from scintillating processes[128]. The former effect allows measuring the dose deposition in the 0.01 to 1 Gy range while the latter is suited for high-bandwidth dose-rate measurements (typically 1 kHz down to 1 mGy/h). For OSL the dose measurement relies on the optical de-excitation of charge carriers previously trapped on energy levels located in the band gap of the OSL crystal. The de-excitation of these charge carriers is induced by laser illumination providing the necessary optical stimulation at the appropriate energy. In the case of scintillation the ionizing radiation directly excites atoms or molecules of a scintillating crystal which then de-excite via radiative transitions. These fiber radiation monitors have recently been evaluated for potential use in large scale nuclear facilities.

8. Concluding Remarks

In this chapter we tried to illustrate the particular challenges of using optical fiber components and sensors in different application fields where the presence of highly energetic radiation is a concern. The radiation tolerance of various optical fiber components has been investigated and by now several component types have been implemented in either fiber communication or sensing systems in space, military, high energy physics and civil nuclear industry applications. All these environments come with very strict reliability requirements that do not provide any legroom for design uncertainties. Therefore and in spite of the consensus on the possible advantages and benefits that photonic systems can bring, many issues still need to be solved before the widespread use of photonic devices in radiation environments will become a reality.

Acknowledgements

The authors wish to acknowledge partial financial support from the European Commission-EURATOM Fusion Technology Program. The content of the publication is the sole responsibility of its publishers and it does not necessarily represent the views of the Commission or its services. Many thoughts brought forward in the text have been shared during discussions with lots of people in the radiation effects field. Particular words of gratitude go to Dr. Edward Taylor (International Photonics Consultants, Inc., USA), Dr. Joe Friebele (Naval Research Laboratory, USA) and Dr. Karl Gill (CERN, Switzerland).

References

1.   A. Holmes-Siedle and L. Adams, *Handbook of radiation effects* (Oxford University Press, Oxford, 1993).
2.   F. Wrobel, in: *Conférence RADECS 2005, Short Course Notebook – New challenges for Radiation Tolerance Assessment,* edited by A. Fernandez Fernandez (Cap d'Agde, 2005), pp. 5–31.
3.   D.L. Griscom, Nature of defects and defect generation in optical glasses, *SPIE Proceedings* 541, 38–59 (1985).
4.   ASTM E722-04e1 Standard Practice for Characterizing Neutron Energy Fluence Spectra in Terms of an Equivalent Monoenergetic Neutron Fluence for Radiation-Hardness Testing of Electronics (ASTM, 2004).
5.   A. Vasilescu and G. Lindstroem, Displacement damage in silicon, on-line compilation (February 2007); http://sesam.desy.de/members/gunnar/Si-dfuncs.html
6.   Radiation belt modeling for living with a star (February 2007); http://radbelts.gsfc.nasa.gov
7.   J.L. Barth, C.S. Dyer and E.G. Stassinopoulos, Space, Atmospheric, and Terrestrial Radiation Environments, *IEEE Transactions on Nuclear Science* 50, 466–482 (2003).
8.   Radiation effects and Analysis, NASA Goddard Space Flight Center (February 2007); http://radhome.gsfc.nasa.gov
9.   A.H. Johnston and B.G. Rax, Proton damage in linear and digital optocouplers, *IEEE Transactions on Nuclear Science* 47, 675–681 (2000).

10. J.J. Jimenez, M.T. Alvarez, R. Tamayo, J.M. Oter, J.A. Dominguez, I. Arruego, J. Sanchez-Paramo and H. Guerrero, Proton radiation effects in high power LEDs and IREDs for optical wireless links for intra-satellite communications, *2006 IEEE Radiation Effects Data Workshop, Workshop Record*, 77–84 (IEEE, 2006).

11. E.J. Friebele and L.R. Wasserman, Development of Radiation-Hard Fiber for IFOGs, *18th International Optical Fiber Sensors Conference Technical Digest*, ME2 (Optical Society of America, Washington DC, 2006).

12. V.M.N. Passaro and M.N. Armenise, Neutron and gamma radiation effects in proton exchanged optical waveguides, *Optics Express* 10, 960–964 (2002).

13. R.S. Fielder, D. Klemer and K.L. Stinson-Bagby, High neutron fluence survivability testing of advanced fiber Bragg grating sensors, AIP Conference Proceedings 699, 650–657 (2004).

14. J. Juergens and G. Adamovsky, Performance Evaluation of Fiber Bragg Gratings at Elevated Temperatures, NASA Report NASA/TM—2004-212888 (NASA, Glenn Research Center, 2004).

15. A.I. Gusarov, D.B. Doyle, N. Karafolas and F. Berghmans, Fiber Bragg gratings as a candidate technology for satellite optical communication payloads: radiation-induced spectral effects, *SPIE Proceedings* 4134, 253–260 (2000).

16. M. Ott and P. Friedberg, Technology Validation of Optical Fiber Cables for Space Flight Environments, *SPIE Proceedings* 4216, 206–217 (2001).

17. M. Ott, Fiber Laser Components, Technology Readiness Overview, *NASA Electronic Parts and Packaging Program, Electronic Parts Project Report* (NASA, 2003).

18. Photonics for Space Environments, *SPIE Proceedings* 1953, edited by E.W. Taylor (1993).

19. Photonics for Space Environments II, *SPIE Proceedings* 2215, edited by E.W. Taylor (1994).

20. Photonics for Space Environments III, *SPIE Proceedings* 2482, edited by E.W. Taylor (1995).

21. Photonics for Space Environments IV, *SPIE Proceedings* 2811, edited by E.W. Taylor (1996).

22. Photonics for Space Environments V, *SPIE Proceedings* 3124, edited by E.W. Taylor (1997).

23. Photonics for Space Environments VI, *SPIE Proceedings* 3440, edited by E.W. Taylor (1998).

24. Photonics for Space and Radiation Environments, *SPIE Proceedings* 3872, edited by E.W. Taylor and F. Berghmans (1999).

25. Photonics for Space Environments VII, *SPIE Proceedings* 4134, edited by E.W. Taylor (2000).

26. Photonics for Space and Radiation Environments II, *SPIE Proceedings* 4547, edited by E.W. Taylor and F. Berghmans (2001).

27. Photonics for Space Environments VIII, *SPIE Proceedings* 4823, edited by E.W. Taylor (2002).

28. Photonics for Space Environments IX, *SPIE Proceedings* 5554, edited by E.W. Taylor (2004).

29. Photonics for Space Environments X, *SPIE Proceedings* 5897, edited by E.W. Taylor (2005).

30. Photonics for Space Environments XI, *SPIE Proceedings* 6308, edited by E.W. Taylor (2006).

31. H. Henschel, M. Körfer, J. Kuhnhenn, U. Weinand and F. Wulf, Fibre optic radiation sensor systems for particle accelerators, *Nuclear Instruments and Methods in Physics Research Section A* 526, 537–550 (2004).

32. CMS Outreach (February 2007); http://cmsinfo.cern.ch/outreach/

33. The ATLAS Experiment (February 2007); http://atlas.ch/

34. D. Inaudi, B. Glisic, S. Fakra, J. Billan, S. Redaelli, J.G. Perez and W. Scandale, Development of a displacement sensor for the CERN-LHC superconducting cryodipoles, *Measurement Science and Technology* 12, 887–896 (2001).

35. Optical links for CMS (February 2007); http://cms-tk-opto.web.cern.ch/

36. R. Macias, M. Axer, S. Dris, K. Gill, R. Grabit, E. Noah, J. Troska and F. Vasey, Advance validation of radiation hardness and reliability of lasers for CMS optical links, *IEEE Transactions on Nuclear Science* 52, 1488–1496 (2005).

37. K. Gill, in: *Conférence RADECS 2005, Short Course Notebook – New challenges for Radiation Tolerance Assessment,* edited by A. Fernandez Fernandez (Cap d'Agde, 2005), 173–219.

38. ITER (February 2007); http://www.iter.org

39. Laser Mégajoule (February 2007); http://www-lmj.cea.fr/html/cea.htm

40. C. Ingesson and J. Palmer, in: *Conférence RADECS 2005, Short Course Notebook – New challenges for Radiation Tolerance Assessment,* edited by A. Fernandez Fernandez (Cap d'Agde, 2005), 139–150.

41. J.L. Bourgade, V. Allouche, J. Baggio, C. Bayer, F. Bonneau, C. Chollet, S. Darbon, L. Disdier, D. Gontier, M. Houry, H.P. Jacquet, J.P. Jadaud, J.L. Leray, I. Masclet-Gobin, J.P. Negre, J. Raimbourg, B. Villette, I. Bertron, J.M. Chevalier, J.M. Favier, J. Gazave, J.C. Gomme, F. Malaise, J.P. Seaux, V. Yu Glebov, P. Jaanimagi, C. Stoeckl, T.C. Sangster, G. Pien, R.A. Lerche and E.R. Hodgson, New constraints for plasma diagnostics development due to the harsh environment of MJ class lasers, *Review of Scientific Instruments* 75, 4204–4212 (2004).

42. Optical fibre sensing and systems in nuclear environments, *SPIE Proceedings* 2425, edited by F. Berghmans and M. Decréton (1994).

43. E. Morange, Capteurs à fibres optiques et réseaux associés, Technical Note ENM/EL 92.94 (Electricité de France, 1992).

44. P. Ferdinand, S. Magne, V. Marty, S. Rougeault, P. Bernage, M. Douay, E. Fertein, F. Lahoreau, P. Niay, J.F. Bayon, T. Georges and M. Monerie, Optical fibre Bragg grating sensors for structure monitoring within the nuclear power plants, *SPIE Proceedings* 2425, 11–20 (1994).

45. J.W. Berthold III, Overview of prototype fiber optic sensors for future application in nuclear environments, *SPIE Proceedings* 2425, 74–83 (1994).

46. A.F. Fernandez, A.I. Gusarov, B. Brichard, S. Bodart, K. Lammens, F. Berghmans, M.C. Decreton, P. Megret, M. Blondel and A. Delchambre, Temperature monitoring of nuclear reactor cores with multiplexed fiber Bragg grating sensors, *Optical Engineering* 41, 1246–1254 (2002).

47. A. Kimura, E. takada, K. Fujita, M. Nakazawa, H. Takahashi and S. Ichige, Application of a Raman distributed temperature sensor to the experimental fast reactor joyo with correction techniques, *Measurement Science and Technology* 12, 966–973 (2001).

48. M. Van Uffelen, F. Berghmans, B. Brichard, P. Borgermans and M.C. Decreton, Feasibility study for distributed dose monitoring in ionizing radiation environments with standard and custom-made optical fibers, *SPIE Proceedings* 4823, 231–221 (2002).

49. T. Kakuta and H. Yaqi, Irradiation tests of electronic components and materials, 3rd International Workshop on Future Electron Devices, RDA/FED, Tokyo, 1986.

50. A. Homes-Siedle, Radiation effects in space, nuclear power and accelerators : impact on optics and light sensors, *SPIE Critical Reviews of Optical Science and Technology* CR66, 37–57 (1997).

51. A. Fernandez Fernandez, B. Brichard, and F. Berghmans, Irradiation facilities at SCK·CEN for radiation tolerance assessment of space materials, *Proceedings ESA symposium materials in a space environment* SP–540 (ESA, 2003).

52. Irradiation of electronic components at Louvain la Neuve (February 2007); http://www.cyc.ucl.ac.be/

53. Radiation effects facility RADEF (February 2007); http://www.phys.jyu.fi/RADEF/main.html

54. ESA/SCC Basic Specification No. 22900 Issue 4, Total dose steady-state irradiation test method (ESA, 1995).

55. TIA-455-64 FOTP-64 Procedure for Measuring Radiation-Induced Attenuation in Optical Fibers and Optical Cables (TIA, 1998).

56. IEC 60793-1-54 Ed 1.0: Optical Fibres – Part 1-54: Measurement methods and test procedures – Gamma irradiation (IEC, 2003).

57. MIL-STD-883E Test method standard – Microcircuits (USA Department of Defense, 1996)

58. E.W. Taylor, Advancement of radiation effects research in photonic technologies: application to space platforms and systems, *SPIE Critical Reviews of Optical Science and Technology* CR66, 58–92 (1997).

59. F. Berghmans, O. Deparis, S. Coenen, M. Decréton and P. Jucker, in: *Trends in Optical Fibre Metrology and Standards*, edited by O.D.D. Soares, *NATO ASI Series E: Applied Sciences* 285, 131–156 (1995).

60. H. Henschel, Radiation hardness of present optical fibres, *SPIE Proceedings* 2425, 21–31 (1994).

61. R.T. Williams and E.J. Friebele, in: *CRC Handbook of laser science and technology III – Optical Materials*, edited by M.J. Weber (CRC press Inc., 1986), pp. 299–499.

62. B. Brichard and A. Fernandez Fernandez, in: *Conférence RADECS 2005, Short Course Notebook – New challenges for Radiation Tolerance Assessment*, edited by A. Fernandez Fernandez (Cap d'Agde, 2005), pp. 95–137.

63. M.N. Ott, Radiation Effects Data on Commercially Available Optical Fiber: Database Summary, *2002 IEEE Radiation Effects Data Workshop*, Workshop Record, 24–31 (IEEE, 2002).

64. D.L. Griscom, M.E. Gingerich and E.J. Friebele, Radiation induced defects in glasses: origin of power-law dependence of concentration on dose, *Physical Review Letters* 71, 1019–1022 (1993).

65. E.J. Friebele, C.G. Askins, C.M. Shaw, M.E. Gingerich, C.C. Harrington, D.L. Griscom, T.E. Tsai, U.C. Paek and W.H. Schmidt, Correlation of single-mode fibre radiation response and fabrication parameters, *Applied Optics* 30, 1944–1957 (1991).

66. D.L. Griscom, Fractal kinetics of radiation-induced point-defect formation and decay in amorphous insulators: Application to color centers in silica-based optical fibers, *Physical Review B* 64, 174201 (2001).

67. Y. Morita and W. Kawakami, Dose rate effects on radiation induced attenuation of pure silica core optical fibers, *IEEE Transactions on Nuclear Science* 36, 584–590 (1989).

68. P. Borgermans and B. Brichard, Kinetic Models and Spectral Dependencies of the Radiation-Induced Attenuation in Pure Silica Fibers, *IEEE Transactions on Nuclear Science* 49, 1439–1445 (2002).

69. M.J. LuValle, E.J. Friebele, F.V. Dimarcello, G.A. Miller, E.M. Monberg, L.R. Wasserman, P.W. Wisk, M.F. Yan and E.M. Birtch, Radiation-Induced Loss

Predictions for Pure Silica Core, Polarization-Maintaining Fibers, *SPIE Proceedings* 6193, 61930J (2006).

70. A. Fernandez Fernandez, B. Brichard, Member and F. Berghmans, *In Situ* Measurement of Refractive Index Changes Induced by Gamma Radiation in Germanosilicate Fibers, *IEEE Photonics Technology Letters* 15, 1428–1430 (2003).

71. W.N. MacPherson, R.R.J. Maier, J.S. Barton, J.D.C. Jones, A. Fernandez Fernandez, B. Brichard, F. Berghmans, J.C. Knight, P.S. Russel and L. Farr, Dispersion and refractive index measurement for Ge, B-Ge doped and photonic crystal fibre following irradiation at MGy levels, *Measurement Science and Technology* 15, 1659–1664 (2004).

72. D.L. Griscom, Optical Properties and Structure of Defects in Silica Glass, *Jounal of the Ceramic Society of Japan* 99, 899–916 (1991).

73. S. Munekuni, T. Yamanaka, Y. Shimogaichi, R. Tohmon, Y. Ohki, K. Nagasawa and Y. Hama, Various types of nonbridging oxygen hole center in high-purity silica glass, *Journal of Applied Physics* 68, 1212–1217 (1990).

74. L. Skuja, K. Kajihara, Y. Ikuta, M. Hirano and H. Hosono, Urbach absorption edge of silica: reduction of glassy disorder by fluorine doping, *Journal of Non-Crystalline Solids* 345 & 346, 328–331 (2004).

75. K. Nagasawa, Y. Hoshi, Y. Ohki and K. Yahagi, Improvement of radiation resistance of pure silica core fibers by hydrogen treatment, *Japanese Journal of Applied Physics* 24, 1224–1228 (1985).

76. B. Brichard, A. Fernandez Fernandez, H. Ooms, F. Berghmans, M. Decréton, A. Tomashuk, S. Klyamkin, M. Zabezhailov, I. Nikolin, V. Bogatyrjov, E. Hodgson, T. Kakuta, T. Shikama, T. Nishitani, A. Costley and G. Vayakis, Radiation-Hardening Techniques Of Dedicated Optical Fibers Used In Plasma Diagnostic Systems In ITER, *Journal of Nuclear Materials*, 329–333, 1456–1460 (2004).

77. D.L. Griscom, Self-trapped holes in amorphous silicon dioxide, *Physical Review B* 40, 4224–4227 (1989).

78. D.L. Griscom, Radiation hardening of pure-silica-core optical fibers: reduction of induced absorption bands associated with self-trapped holes, *Applied Physics Letters* 71, 175–177 (1997).

79. B. Brichard, P. Borgermans, A. Fernandez Fernandez, K. Lammens and M. Decréton, Radiation Effect in Silica Optical Fibre exposed to intense mixed neutron-gamma radiation field, *IEEE Transactions on Nuclear Science* 48, 2069–2073 (2001).

80. B. Brichard, A. Fernandez Fernandez, F. Berghmans and M. Decréton, Origin of the radiation-induced OH vibration band in polymer-coated optical fibres irradiated in a nuclear fission reactor, *IEEE Transactions on Nuclear Science* 49, 2852–2856 (2002).

81. D.L. Griscom, M.E. Gingerich, and J.E. Friebele, Model for the dose, dose-rate and temperature dependence of radiation-induced loss in optical fiber, *IEEE Transactions on Nuclear Science* 41, 523–527 (1994).

82. J.E. Golob, P.B. Lyons and L.D. Looney, Transient radiation effects in low-loss optical waveguides, *IEEE Transactions on Nuclear Science* 24, 2164–2168 (1977).

83. L.D. Looney, and P.B. Lyons, Radiation-induced transient absorption in single mode optical fibers, *SPIE Proceedings* 992, 84–91 (1988).

84. E.J. Friebele, P.B. Lyons, J. Blackburn, H. Henschel, E.W. Taylor, G.T. Beauregard, R. H. West, P. Zagarino, and D. Smith, Interlaboratory comparison of radiation-induced attenuation in optical fibers: part III: transient exposures, *IEEE Journal of Lightwave Technology* 8, 977–989 (1990).

85. S. Girard, J. Baggio, J-L. Leray, J-P. Meunier, A. Boukenter, and Y. Ouerdane, Vulnerability analysis of optical fibers for Laser Megajoule facility: preliminary studies, *IEEE Transactions on Nuclear Science* 52, 1497–1503 (2005).

86. S. Girard, J. Keurinck, Y. Ouerdane, J-P. Meunier, and A. Boukenter, Gamma-rays and pulsed X-ray radiation responses of germanosilicate single-mode optical fibers: influence of cladding codopants, *IEEE Journal of Lightwave Technology* 22, 1915–1922 (2004).

87. L.D. Looney, P.B. Lyons, W. Schneider, and H. Henschel, Influence of preform variations and drawing conditions on transient radiation effects in pure silica fibers, *SPIE Proceedings* 721, 37–43 (1986).

88. S. Girard, Y. Ouerdane, A. Boukenter, and J-P. Meunier, Transient radiation responses of silica-based optical fibers: influence of modified chemical-vapor deposition process parameters, *Journal of Applied Physics* 99, 0231041-5 (2006).

89. S. Girard, D.L. Griscom, J. Baggio, B. Brichard, and F. Berghmans, Transient optical absorption in pulsed-X-ray-irradiated pure-silica-core optical fibers: influence of self-trapped holes, *Journal of Non-Crystalline Solids* 352, 2637–2642 (2006).

90. K. Tanimura, C. Itoh, and N. Itoh, Transient optical absorption and luminescence induced by band-to-band excitation in amorphous SiO_2, *Journal of Physics C Solid State Physics* 21, 1869–1876 (1988).

91. S. Girard, A. Yahia, A. Boukenter, Y. Ouerdane, J.-P. Meunier, R.E. Kristiaensen and G. Vienne, g-radiation induced attenuation in photonic crystal fibre, *Electronics Letters* 38, 1169–1170 (2002).

92. A.F. Kosolapov, S.L. Semjonov and A.L. Tomashuk, Improvement of radiation resistance of multimode silica-core holey fibers, *SPIE Proceedings* 6193, 61931E (2006).

93. S. Girard, J. Baggio, and J.-L. Leray, Radiation-Induced Effects in a New Class of Optical Waveguides: The Air-Guiding Photonic Crystal Fibers, *IEEE Transactions on Nuclear Science* 52, 2683–2688 (2005).

94. H. Henschel, J. Kuhnhenn and U. Weinand, High radiation hardness of a hollow core photonic bandgap fiber, *RADECS 2005 Conference Proceedings* LN4 (2005).

95. G. Messenger and J. Spratt, The effect of neutron irradiation on silicon and germanium, *Proceedings of the Institute of Radio Engineers* 46, 1038–1044 (1958).

96. B.H. Rose and C. Barnes, Proton damage effects on light emitting diodes, *Journal of Applied Physics* 53, 1772–1780 (1982).

97. A.H. Johnston, Radiation Effects in Light-Emitting and Laser Diodes, *IEEE Transactions on Nuclear Science* 50, 689–703 (2003).

98. P. Le Metayer, O. Gilard, R. Germanicus, D. Campillo, F. Ledu, J. Cazes, W. Falo and C. Chatry, Proton damage effects on GaAs/GaAlAs veritcal cavity surface emitting lasers, *Journal of Applied Physics* 94, 7757–7763 (2003).

99. Y.F. Zhao, A.R. Patwary, R.D. Shrimpf, M.A. Netfeld and K.F. Galloway, 200 MeV proton damage effects on multi-quantum well laser diodes, *IEEE Transactions on Nuclear Science* 44, 1898–1905 (1997).

100. R. Macias, M. Axer, S. Dris, K. Gill, R. Grabit, E. Noah, J. Troska, and F. Vasey, Advance validation of radiation hardness and reliability of lasers for CMS optical links, *IEEE Transactions on Nuclear Science* 52, 1488–1496 (2005).

101. F. Berghmans, M. Van Uffelen and M. Decréton, High total dose gamma and neutron radiation tolerance of VCSEL assemblies, *SPIE Proceedings* 4823, 162–171 (2002).

102. M. Van Uffelen, J. Mols, F. Berghmans, Ionizing radiation assessment of long-wavelength VCSELs up to MGy dose levels, *RADECS 2006 Workshop Proceedings,* PG-1 (2006).

103. A. Kalavagunta, B. Choi, M.A. Neifeld and R. Shrimpf, Effects of 2 MeV proton irradiation on operating wavelength and leakage current of vertical cavity surface emitting lasers, *IEEE Transactions on Nuclear Science* 50, 1982–1990 (2003).

104. J. Troska, K. Gill, R. Grabit and F. Vasey, Neutron, proton and gamma radiation effects in candidate InGaAs p-i-n photodiodes for the CMS tracker optical links, *Tech. Rep. No. CMS-NOTE-1997-102* (CERN, 1997).

105. S. Onoda, Spectral response of a gamma and electron irradiated pin photodiode, *IEEE Transactions on Nuclear Science* 49 1446–1449 (2002).

106. K. Gill, M. Axer, S. Dris, R. Grabit, R. Macias, E. Noah, J. Troska, and F. Vasey, Radiation Hardness Assurance and Reliability Testing of InGaAs Photodiodes for Optical Control Links for the CMS Experiment, *IEEE Transactions on Nuclear Science* 52, 1480–1487 (2005).

107. M. Van Uffelen, I. Genchev and F. Berghmans, Reliability study of photodiodes for their potential use in future fusion reactor environments, *SPIE Proceedings* 5465, 92–102 (2004).

108. F. Berghmans, F. Vos and M. Decréton, Evaluation of three different optical fibre temperature sensor types for application in gamma-radiation environment, *IEEE Transactions on Nuclear Science* 45, 1537–1542 (1998).

109. K. Krebber, H. Henschel and U. Weinand, Fibre Bragg gratings as high dose radiation sensors?, *SPIE Proceedings* 5855, 176–179 (2005).

110. P. Niay, P. Bernage, M. Douay, F. Lahoreau, J. Bayon, T. Georges, M. Monerie, P. Ferdinand, S. Rougeault and P. Cetier, Behaviour of Bragg gratings, written in germanosilicate fibers, against γ-ray exposure at low dose rate, *IEEE Photonics Technology Letters* 6, 1350–1352 (1994).

111. S.A. Vasiliev, E.M. Dianov, K.M. Golant, O.I. Medvedkov, A.L. Tomashuk, V.I. Karpov, M.V. Grecov, A.S. Kurkov, B. Leconte, et al., Performance of Bragg and long-period gratings written in N- and Ge-doped silica fibers under γ-radiation, *IEEE Transactions on Nuclear Science* 45, 1580–1583 (1998).

112. A.I. Gusarov, F. Berghmans, O. Deparis, A. Fernandez Fernandez, Y. Defosse, P. Mégret, M. Decréton, and M. Blondel, High total dose radiation effects on temperature sensing fibre Bragg gratings, *IEEE Photonics Technology Letters* 11, 1159–1161 (1999).

113. A.I. Gusarov, F. Berghmans, A. Fernandez Fernandez, O. Deparis, Y. Defosse, D. Starodubov, P. Mégret, M. Décreton, and M. Blondel, Behaviour of fibre Bragg gratings under high total dose gamma radiation, *IEEE Transactions on Nuclear Science* 47, 688–692 (2000).

114. A.I. Gusarov, A. Fernandez Fernandez, F. Berghmans, S.A. Vasiliev, O. Medvedkov, M. Decréton, O. Deparis, P. Mégret and M. Blondel, Effect of combined gamma-neutron radiation of multiplexed fiber Bragg grating sensors, *SPIE Proceedings* 4134, 86–95 (2000).

115. T.-E. Tsai, G.M. Williams and E.J. Friebele, Index structure of fiber Bragg gratings in Ge-SiO2 fibers, *Optics Letters* 22, 224–226 (1997).

116. A. Fernandez Fernandez, B. Brichard, F. Berghmans, and M. Decréton, Dose-rate dependencies in gamma-irradiated Fiber Bragg grating filters, *IEEE Transactions on Nuclear Science* 46, 2874–2878 (2002).

117. E.W. Taylor, K. Hulick, J.M. Battiato, A.D. Sanchez, J.E. Winter and A. Pirich, Response of germania doped fiber Bragg gratings in radiation environments, *SPIE Proceedings* 3714, 106–113 (1999).

118. A. Fernandez Fernandez, A. Gusarov, B. Brichard, M. Decréton, F. Berghmans, P. Mégret and A. Delchambre, Long-term radiation effects on fibre Bragg grating temperature sensors in a low flux nuclear reactor, *Measurement Science and Technology* 15, 1506–1511 (2004).

119. S.A. McElhaney, D.D. Falter, R.A. Todd, M.L. Simpson, J.T. Mihalczo, Passive (self-powered) fiber optic sensors, *Conference Record of the 1992 IEEE Nuclear Science Symposium and Medical Imaging Conference* 1, 101–103 (1992).

120. C. Lai, W. Lee and W. Wang, Gamma Radiation Effect on the Fiber Fabry–Pérot Interference Sensor, *IEEE Photonics Technology Letters* 15, 1132–1134 (2003).

121. A. Fernandez Fernandez, P. Rodeghiero, B. Brichard, F. Berghmans, A. H. Hartog, P. Hughes, K. Williams and A.P. Leach, Radiation-Tolerant Raman Distributed Temperature Monitoring System for Large Nuclear Infrastructures, *IEEE Transactions on Nuclear Science* 52, 2689–2694 (2005).

122. D. Alasia, A. Fernandez Fernandez, L. Abrardi, B. Brichard and L. Thévenaz, The effects of gamma-radiation on the properties of Brillouin scattering in standard Ge-doped optical fibres, *Measurement Science and Technology* 17, 1091–1094 (2006).

123. H. Bueker, F.W. Haesing, and E. Gerhard, Physical properties and concepts for applications of attenuation-based fiber optic dosimeters for medical instrumentation, *SPIE Proceedings* 1648, 63–70 (1992).

124. M.C. Decréton, V. Massaut and P. Borgermans, Potential benefit of fibre optics in nuclear applications – the case of decommissioning and storage activities, *SPIE Proceedings* 2425, 2–10 (1994).

125. B. Brichard, P. Borgermans, F. Berghmans, M. Decréton, A.L. Tomashuk, I.V. Nikolin, R.R. Khrapko and K.M Golant, Dedicated optical fibres for dosimetry based on radiation-induced attenuation : experimental results, *SPIE Proceedings* 3872, 36–42 (1992).

126. A. Fernandez Fernandez, S. O'Keeffe, C. Fitzpatrick, B. Brichard, F. Berghmans and E. Lewis, Gamma dosimetry using commercial PMMA optical fibres for nuclear environments, *SPIE Proceedings* 5855, 499–502 (2005).

127. L. Dusseau and J. Gasiot, Online and realtime dosimetry using optically stimulated luminescence, *International Journal of High Speed Electronics and Systems* 14, 605–623 (2004).

128. A.L. Huston, B.L. Justus, P.L. Falkenstein, R.W. Miller, H. Ning and R. Altemus, Remote optical fiber dosimetry, *Nuclear Instruments and Methods in Physics Research B* 184, 55–67 (2001).

SILICON-BASED MICROPHOTONICS FOR BIOSENSING APPLICATIONS

S. JANZ, A. DENSMORE, D.-X. XU, P. WALDRON,
A. DELÂGE, P. CHEBEN, J. LAPOINTE, J.H. SCHMID
*Institute for Microstructural Sciences, National Research
Council Canada, Ottawa, Ontario, Canada, K1J 6B1*

Abstract. This paper reviews theory and experiments on silicon photonic wire waveguide evanescent field (PWEF) biosensors. Theoretical considerations and supporting calculations show that sensor response increases both with increasing core-cladding refractive index contrast, and with decreasing waveguide core thickness until a maximum sensor response is achieved. As a result, appropriately designed Si waveguide sensors can have the largest response to superstrate refractive index shifts, and also to surface molecular adsorption, of any commonly available waveguide system. Measurements of Si waveguide sensor response to fluid index change and biotin-avidin binding reactions confirm the predictions of theory.

Keywords: Evanescent field; biosensors; waveguides; silicon-on-insulator; photonic wires; silicon photonics

1. Introduction: Free Space and Waveguide Optics

The interaction of light and matter has been and remains the most important source of knowledge and understanding of the physical world around us. In everyday life human beings rely on light to find their way through a complex and changing world. In science also, light has provided the window that connects us to the microscopic world of atoms and molecules and allows us to probe their structure and interactions. As the physical, chemical and biological sciences continue to overlap, ever more sophisticated photonics tools are being developed to probe the energetics and dynamics of molecular processes in real time. In this paper, we focus on

W.J. Bock et al. (eds.), Optical Waveguide Sensing and Imaging, 167–194.
© 2008 *Springer.*

silicon based evanescent field waveguide sensors that use light to monitor the recognition chemistry and kinetics of affinity binding systems.

Light can be used as a probe to extract information in many ways (Figure 1). In the simplest case light is scattered from objects or particles with no change in photon energy or wavelength. This tells us mainly something about the macroscopic (i.e. at length scales comparable to or larger than the wavelength of light) structure of the scattering object. In Raman scattering or photoluminescence, light is scattered with a change in photon energy. The incident photon was absorbed by an available electronic, vibrational, or rotational transition in the molecules or atoms, and re-emitted as a new photon shifted in energy. Since the allowed energy change of these events can be related to the underlying atomic structure of the system, such inelastic scattering processes are much more useful in developing a picture of the atomic scale physics of the system. Finally through transmission and reflection experiments one can measure the linear absorption α and index of refraction n, which are directly related to the dielectric function ε of the material. Both n and α carry information on the density and electronic structure of the material, and can in fact be mathematically transformed from one to the other through the Kramers-Kronig transformations.

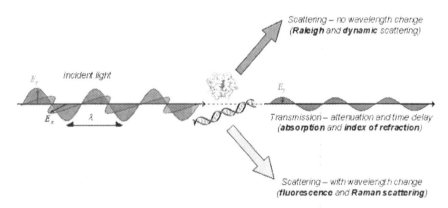

Figure 1. Light interacts with matter through scattering without wavelength change, scattering with wavelength change as in the case of Raman scattering and photolumine-scence, and through absorption and the index of refraction as it propagates through materials.

Optical experiments in the past traditionally employed free space optics. Light is allowed to propagate in free space according to the rules of wave propagation and diffraction, while its direction, beam shape and wavelength discrimination must be determined by the suitable placement of bulk optical elements such as lenses, mirrors and gratings (Figure 2). In free space and bulk optic experiments, diffraction limits the minimum volume element that light can probe to about one wavelength in cross-section, which is approximately one micrometer in the case of visible and near-infrared light. This is much larger than the length scale of even large biological molecules. One consequence is that it is difficult to detect and monitor molecular binding reactions involving a small subset of molecules in a given sample. This is changing with the increasing use of waveguide optics, integrated optics and the emerging nanophotonic devices. Using waveguides it is now possible to arbitrarily specify an optical circuit that determines the destination, optical path, phase and interaction light with a matter at the sub-micrometer scale in a way that was simply not possible three decades ago.

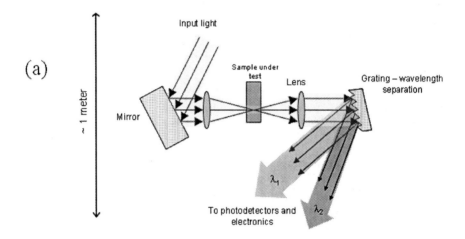

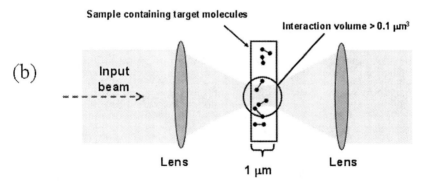

Figure 2. (a) Free space optics relies of the transmission of light according to the laws of free space propagation, and the use of bulk optical elements such as mirrors and lenses to focus and redirect light. (b) The minimum volume of material that light can interact with is fixed by the minimum achievable focal spot size, and the depth of field of the focal spot. For near infrared wavelengths, this corresponds to a volume of 0.1 μm^3.

A waveguide such as the one shown in Figure 3 (a) is formed by a core layer of material with index n_{core} surrounded by material of lower index of refraction. In the simplest picture, when a light ray is launched in the core layer at a sufficiently small angle relative to the core cladding interface, the ray will be confined in the core by total internal reflection, and propagate along the waveguide indefinitely with no loss. A more complete theoretical picture is obtained if one solves Maxwell's equations for the electro-magnetic field in a waveguide. The waveguide core is analogous to an optical potential well that confine light in the lateral direction. The result of such a calculation [1] is a limited set of confined optical modes which can propagate along the waveguide, each with a phase velocity of $v = c/N_{eff}$, where N_{eff} is the effective index of the mode and c is the speed of light in vacuum. All the light confined to the waveguide must propagate as a superposition of one or more of these modes. In the remainder of this paper we assume that the waveguides support only one fundamental mode with effective index N_{eff}. In light of the evanescent filed sensors described below, it should be noted here that N_{eff} is determined by the material refractive indices of both the core and the cladding layers[1].

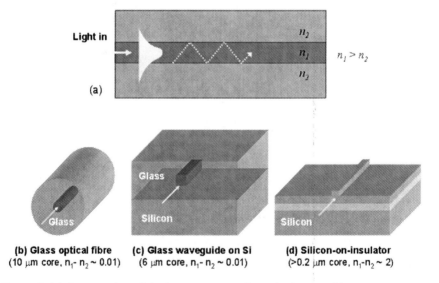

Figure 3. (a) Cross-section of the layer structure of a typical waveguide, and examples of common waveguides including (b) glass optical fiber, (c) a glass waveguide on a silicon substrate, and (d) a silicon-on-insulator ridge waveguide.

The most familiar waveguide is the glass optical fiber in Figure 3(b) formed by a cylindrical core of glass with a typical diameter of 8 μm in a single mode fiber, surrounded by a much thicker glass cladding with a slightly lower index. Here the index step Δn between core and cladding usually less than 1% or about $\Delta n = 0.01$. Figure 3(b) shows a glass waveguide as would be fabricated on a silicon substrate, as would be used in a variety of silica on silicon integrated optical devices, or planar lightwave circuits (PLC). Although the core is rectangular in shape the index step and core dimensions are comparable to those for a glass fiber. Given the small index step, the smallest possible core size of such glass waveguides is several micrometers. If the waveguide core becomes any smaller, the optical mode begins to expand into the surrounding cladding.

2. Silicon Microphotonics

There has been an explosion of activity in the area of silicon photonics over the last five years[2–3]. The Si waveguides can be as small as a few hundred nanometers in cross section; less than the wavelength of light. These silicon photonic wires may be used to build functional optical circuits that are only a few tens of micrometers or less in length, several hundred times smaller than comparable devices on the older glass technologies. Furthermore, the SOI wafer starting material is a readily available and inexpensive commodity used by the electronics industry, and silicon fabrication processes may be used to fabricate devices at the wafer scale. Although this field has been driven by the need for silicon-based optical interconnects, the arising technology will also have great impact on waveguide based optical biosensors.

Silicon photonic wire sensors are fundamentally different from previous evanescent field (EF) waveguide sensor devices. In this paper, we demonstrate through both theory and experimental verification that silicon photonic wire evanescent field (PWEF) sensors have the highest response to a molecular binding event, compared with any other readily available waveguide platform. However the most important practical impact of photonic wire technology may arise from more mundane practical considerations. Individual PWEF sensor elements in SOI can be 100 µm or less in size, comparable to or smaller that the spots on a typical micro-array chip. In comparison, a single glass waveguide EF sensor must be more than a centimeter long to achieve the needed response. Silicon waveguides sensors may also be made long to enhance sensitivity, but because the PWEF sensor waveguides can be folded with extremely small bend radii of 10 µm or smaller, a folded 1 cm long waveguide can still fit into a 100 µm diameter spotter target area. Alternatively, a complete sensor of 10 µm in size can be formed using a ring resonator. As a first step in this direction, in this paper we also demonstrate sensor operation in two simple but compact optical circuits capable of monitoring individual sensor elements continuously in real time: a photonic wire Mach-Zehnder interferometer and a ring resonator.

Silicon waveguides are almost always formed on silicon-on-insulator (SOI). These wafers consist of a conventional silicon substrate wafer, a thin (~ 1 μm) buried oxide (BOx) layer of SiO_2, and finally a top layer of single crystal silicon. This top layer can range in thickness from 0.2 μm to several micrometers, and forms a natural slab waveguide in which light is confined by the buried oxide below, with index n = 1.46, and either air or a similar capping oxide layer above. Lateral optical confinement can be obtained by etching ridges or channels into the top Si layer, so that SOI can used to fabricate integrated optical waveguide circuits in much the same way as for glass and III-V semiconductor integrated optics. Silicon is optically transparent at wavelengths longer than λ=1200 nm, and has an index of refraction of n = 3.5 at λ=1550 nm. Therefore unlike glass waveguides and III-V semiconductor waveguides, the index contrast between the waveguide core and surrounding cladding is of the order of Δn = 2 on all sides of the waveguide. Figure 3(d) shows a silicon-on-insulator (SOI) waveguide, and Figure 4(a) shows a cross-sectional electron microscope image of a fabricated 2 μm thick SOI ridge waveguide. Here the waveguide core is formed in a thin single crystal layer of silicon layer on top of a buried SiO_2 layer. Such high index contrast waveguides have markedly different properties than the glass waveguides. For example it is possible to shrink the waveguide core size down to a few hundred nanometers in cross-section as in Figure 4(b) while keeping light tightly confined to the waveguide. Waveguide device size is often limited by how tightly a waveguide can turn a corner without suffering from excessive optical bend losses. Si photonic wire waveguides can bend with a radius of curvature as small as 10 μm or less, with negligible loss as illustrated by the spiral test structure in Figure 5. The minimum bending radius for glass waveguides is several millimeters. This means that complex optical circuits can be fabricated in SOI with overall dimensions of a few hundred micrometers, whereas the equivalent circuit in glass can take up several square centimeters of wafer area.

The high index contrast in SOI waveguides also provides a unique ability to manipulate light at the sub-wavelength scale. For example SOI is the most commonly used platform for two-dimensional photonic crystals. For application as a PWEF biosensor, the most important feature of the SOI waveguide system is that for thin waveguides, much of the guided mode electrical field is localized to thin layer within 100 nm just above the waveguide surface. As a result the interaction of the guided mode with molecules near the surface can be much stronger than for any other waveguide system.

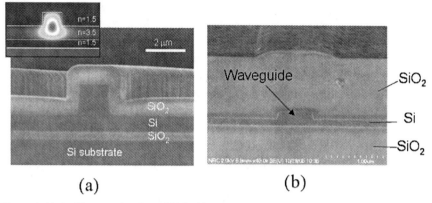

(a)   (b)

Figure 4. (a) A silicon-on-insulator (SOI) ridge waveguide with a 2 μm thick Si guiding layer and SiO₂ oxide top cladding. The inset shows the calculated optical mode profile for this waveguide. (b) A 0.26 × 0.45 μm SOI photonic wire ridge waveguide.

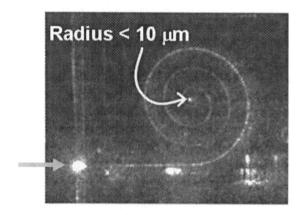

Figure 5. A spiral photonic wire waveguide test structure with minimum bend radius of 10 μm. The waveguide traced by scattered light was imaged using an IR camera while λ =1550 nm light was coupled into the waveguide.

3. Label Free Sensing

The detection and recognition of molecules in genomics, pathogen detection and proteomics is achieved through the use of affinity binding between matched pairs of molecules. Examples of such matched pairs include the two complementary strands of a single DNA molecule, or an antibody-antigen pair. For example, one may test for the presence of a certain DNA molecule extracted from pathogenic bacteria by detecting when a target bacterial DNA strand binds to a previously prepared complementary DNA strand. Since binding of such matched pairs can be highly specific, this mechanism provides a natural and extremely effective solution the problem of molecular recognition. However the difficult challenge facing device scientists lies in detecting a small number of such binding events with little of no false positives. Most current methods involve labeling the target molecules with a fluorescent marker that emits light of a specific wavelength when illuminated. The receptor molecules may be attached to a substrate, and are exposed to a solution containing the sample molecules. If target molecules are present, they will bind to the receptor and remain attached after the sample solution is removed and the chip surface is washed. The presence of the target molecule can then be detected by illuminating the chip with light to excite the fluorescence of the tag. The method can be extremely effective but requires several extra steps in sample preparation to attach the fluorescent tags, and the reading the chips after exposure and washing. It is also not always possible to attach tags to certain molecules.

As a result, there is considerable interest in label-free sensing for which the binding event can be detected without the use of tags, and also where binding reactions can be monitored continuously in real time. A number of different approaches have been explored based on electrical and optical techniques. Electrical techniques generally involve measuring the changes in surface potential when charged molecules bind to the surface, and as a result can only be used to probe for molecules that are charged or have a strong dipole moment. Electrochemical methods are also very sensitive to the pH of solutions used to carry the samples. On the other hand label free optical techniques are all based on measuring the local change in index of refraction induced by the presence of molecules. If optical resonances can be neglected to a first approximation, such index based optical sensors can be regarded as mass sensors since the average index change occurring in a small volume of liquid is proportional to the size of the molecule and the number density of molecules localized within that volume. One approach is to use a Fabry-Perot cavity formed in porous silicon (p-Si)[4]. The sample fluid is allowed to penetrate the p-Si. Molecular binding is signaled by a permanent change in the resonance wavelength after the resonator device has been exposed to the sample and then washed to remove unbound molecules. Another approach measures the increase in reflectivity of a surface with an anti-reflection (AR) coating, as the target molecules bind to the surface.[5]

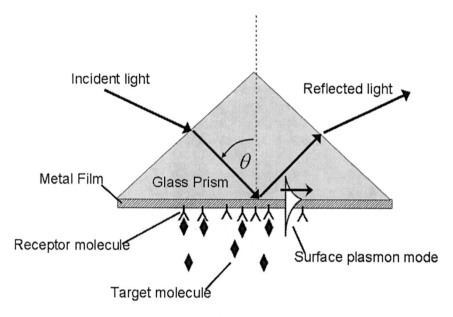

Figure 6. A schematic diagram of a prism coupled surface plasmon resonance sensor.

At this time the only label free affinity sensor that is commercially available and in regular use in research and industrial laboratories is based on surface plasmon resonance (SPR)[6]. The principle of SPR is illustrated in Figure 6. An incident light beam is coupled into a propagating surface plasmon mode in thin metal film deposited on the bottom of a prism. The surface plasmon is a propagating electromagnetic mode that is guided along the metal surface, and is localized to within a few hundred nanometers of the metal surface. When the incident light couples to the surface plasmon mode, a sharp decrease in reflected power is observed. However, coupling can only occur when the component of the incident light wavevector $k_i = n(\omega/c)$ parallel to the surface is equal to the surface-plasmon wavevector. This matching condition requires the use of a prism coupling geometry as shown in Figure 6. Furthermore, wave vector matching only occurs at a specific angle θ for any given incident wavelength. This angle depends on the optical constants and thickness of the metal, and more importantly, on the index of refraction in a region immediately adjacent to the metal surface. In a sensor configuration, the receptor molecules are fixed to the metal film surface. When target molecules attach to these receptors, they change the local refractive index and thereby shift the SPR resonance angle. Thus binding is detected by monitoring the SPR resonance angle conti-nuously during an interaction. SPR sensors are extremely sensitive and can be used to detect as little as a few picograms of molecules per square millimeter area. SPR sensors are essentially bulk optical devices and do not lend themselves well to integration into arrays of multiple sensors. However they are one of the few tools that can be used to monitor binding affinity reactions in real time and while conditions such as temperature and solution composition are varied. This is one reason why SPR systems are often found in research laboratories, since important information on molecular adsorption, desorption and binding strength can be extracted from such experiments.

In the remainder of this paper we discuss a closely related type of sensor, the evanescent field (EF) waveguide sensor.[7–9] EF sensors are very similar to SPR sensors in operating principle, and share the advantage of being able to monitor binding reactions continuously and in real time. The role of the metal film and the plasmon mode in SPR is now played by an optical waveguide and the guided optical mode, with its evanescent field that interacts with molecules on the waveguide surface.

4. Silicon Photonic Wire Evanescent Field Sensors: Theory

An SOI waveguide evanescent field sensor is shown in cross-section in Figure 7. In this device sensing is a result of the interaction of the guided mode evanescent tail interacting with the ambient liquid and molecules adjacent to the waveguide surface. Most of the waveguide circuit is covered with an impermeable dielectric layer which may be SiO_2 or a polymer such as SU-8. The only restriction is that the materials in contact with the sample fluid are compatible with the biological molecules and chemistries used. A window is opened in this cladding layer to allow the sample fluid to be in contact with the waveguide surface. The surface in this sensing window must be functionalized with a receptor molecule chosen to bind specifically to the target molecule or analyte being tested. In the case of genetic tests the receptor can be a single strand of DNA that is complementary to at least a short single strand segment of target DNA. Alternatively, antibodies that bind to target proteins and disease markers of diagnostic value may be attached to the surface. Chemistries for attaching receptor molecules to surface such as glass have been developed for many years and are already in use for commercial fluorescent tag based bio-chips. Research into receptor-analyte pairs appropriate for specific diagnostic or pathogen detection has also been pursued for several decades.

Once the sensor surface has been appropriately functionalized, light is launched into the waveguide core. The waveguide mode propagation is

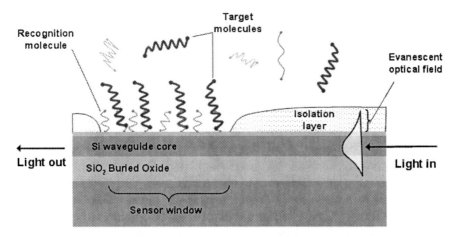

Figure 7. Cross-section of an SOI evanescent field (EF) waveguide sensor.

characterized the effective index N_{eff}. The phase velocity of the guide optical mode is given by $v = c/N_{eff}$, and the propagation wavevector is $\beta = N_{eff}\,(\omega/c)$. As shown in Figure 7, the electric field of the waveguide mode has an evanescent tail that extends outside of the core layer into the upper and lower cladding material. The evanescent tail interacts with the material or liquid immediately above the surface, and the value of N_{eff} will be determined by the index of refraction of both the core layer, and upper and lower claddings. In the sensor window, the index of refraction immediately above the waveguide surface will increase if target molecules bind to the receptor molecules. This will shift the effective index of the waveguide mode by an amount δN_{eff} that depends on the overlap of the modal field distribution with the molecular layer, the density of adsorbed molecules and the size of the molecules. As a result, as molecules bind to the surface the phase of the light propagating through the sensor window will be shifted by an amount

$$\Delta\phi = \delta N_{eff}\left(2\pi/\lambda\right)L. \qquad (1)$$

The optical phase change $\Delta\phi$ provides a transduction signal that can be easily measured using interferometric techniques that will be described in section 5. Using a perturbation theory approach,[10] the change in effective index may be expressed as an overlap integral of the waveguide mode field distribution with the local perturbation of dielectric function $\Delta\varepsilon$ or corresponding material index perturbation Δn.

$$\delta N_{eff} = c \int \Delta \varepsilon \, E_0 \, E_0^* \, dy \, dx$$
$$= c \int 2n \, \Delta n \, E_0 \, E_0^* \, dy \, dx$$

(2)

Here x and y are the coordinates in the plane perpendicular to the wave-guide propagation direction. Equation 1 shows that increasing the response of the waveguide sensor to molecular binding events can be accomplished by increasing the length L of the sensor window, albeit at the cost of increasing sensor size. Although increasing length increases the response to molecular density adsorbed on the waveguide surface, the response to the total number of molecules adsorbed remains the same. On the other hand, Equation 2 shows that the sensor response per unit length can be improved by increasing the fractional overlap of the adsorbed molecular layer with the waveguide mode field amplitude distribution. This optimization path increases the absolute sensitivity to the total number of adsorbed molecules.

To increase the mode overlap with a surface adsorbed layer, one must increase the fraction of the mode power that is contained in the evanescent field outside the waveguide core in the upper cladding. This is easily achieved by either reducing the index step between the waveguide core layer and upper cladding layer, or by decreasing the thickness of the waveguide core. Both approaches reduce the confinement of the waveguide mode in the waveguide core. For example, in glass waveguides with an index step of approximately 1%, reducing the waveguide core thickness below about 3 μm does not result in a corresponding decrease in mode size. Instead, a larger fraction of the field propagates in the cladding layers. If the core thickness is further reduced, the mode actually begins to expand into the cladding. In SOI waveguides, the high index contrast is much more effective in confining the field to the waveguide core, which can be reduced to dimensions less than 0.5 μm before a significant fraction of the waveguide mode propagates in the cladding region.

Even if much of the modal power propagates in the cladding regions, if the evanescent field decays slowly from the waveguide surface and extends a large distance into the upper cladding, little improvement in adsorbed molecule-modal field overlap is achieved. Therefore one must also improve the localization of the evanescent field to the adsorbed layer that typically extends only a few tens of nanometers from the waveguide surface. The evanescent field decays exponentially into the cladding as exp(-κy) with a decay constant κ that depends on the difference between the mode effective index and the index of refraction of the cladding layer [1].

$$\kappa = \left(\frac{2\pi}{\lambda}\right)\sqrt{N_{eff}^2 - n_{clad}^2} \tag{3}$$

This simple result predicts that strong electric field localization in the near surface region requires a waveguide with as high an N_{eff} as possible. For a given core thickness this implies that the waveguide core should have the highest possible material refractive index.

To summarize the above discussion, it is now clear that the optimal evanescent field sensor waveguide should be (i) extremely thin to ensure that most of the mode power propagates in the cladding, and (ii) have the highest possible core refractive index to concentrate the evanescent field at the waveguide surface where the molecular binding interactions taking place. A survey of the commonly available waveguide systems suggests that thin SOI waveguides are therefore the optimal choice for fabricating EF waveguide sensor devices. Silicon waveguides fabricated on the SOI platform can be designed with cross-sectional dimensions as small as 0.2 μm. Such waveguides have recently been referred to as photonic wire waveguides, and have been used to demonstrate a number of extremely compact integrated optical devices and circuits.[11-13]

There is one further effect which contributes to the high response of SOI based EF sensors. Electromagnetic boundary conditions must be satisfied at the waveguide surface. First the tangential component of the electric field vector **E** must be continuous across the core-cladding interface, while the perpendicular component of the displacement $\boldsymbol{D} = \varepsilon\boldsymbol{E}$ must be continuous. The latter condition demands that the perpendicular electric field have a discontinuous step across the core-cladding boundary located at y, such that the ratio of the field just outside the core $E(y^+)$ to that just below the surface $E(y^-)$ is inversely proportional to the ratio of the core and cladding dielectric constants.

$$\frac{E(y^+)}{E(y^-)} = \frac{\varepsilon_{core}}{\varepsilon_{clad}} = \left(\frac{n_{core}}{n_{clad}}\right)^2 \tag{4}$$

As before, Eq. 4 suggests that the choosing a waveguide system with a high core refractive index will enhance modal electric field $E(y^+)$ in the near surface region of the sensor waveguide. In this case, the polarization of the waveguide mode must be the transverse magnetic (TM) mode which has the dominant field component directed normal to the waveguide surface.

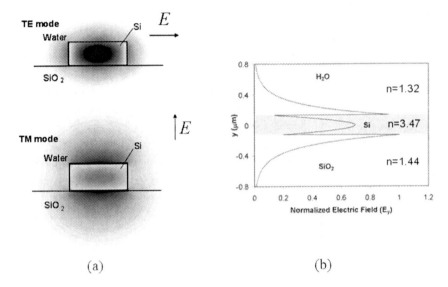

(a)   (b)

Figure 8. (a) TE and TM mode profiles for a 0.22 × 0.45 µm Si photonic wire waveguide.
(b) The electric field amplitude along the center line of the waveguide for the TM mode.

Figure 8 shows the shape of the calculated TM-like and TE-like
waveguide modes for a 0.22 µm thick SOI photonic wire waveguide. The
calculations were carried out using a fully vectorial finite difference method.
The results clearly illustrate the difference between the TE and TM polarized
modes arising from the electromagnetic boundary conditions. For the TM
mode more than half the modal power is propagating in the cladding, yet
the evanescent field remains localized to with about 100 nm of the surface;
precisely what is needed for an optimized evanescent field waveguide
sensor.

Further calculations were carried out to determine the response of an
SOI PWEF sensor to adsorption of a molecular monolayer, optimize the
waveguide design, and also compare the EF sensor response for several
different waveguide types. In these simulations the molecular monolayer
was represented by a 4 nm thick layer with an index of refraction of $n = 1.5$.
The ambient upper cladding was assumed to be water with a refractive
index of $n = 1.32$. For all examples, the lower cladding was assumed to be
SiO_2 with and index of $n = 1.46$. In the simulations, the difference in
effective index δN_{eff} induced by the addition of the 4 nm molecular adlayer
was calculated as a function of the core waveguide thickness. The
wavelength in all cases was assumed to be $\lambda = 1550$ nm. The results of
these simulations are shown in Figure 9 for glass ($n_{core} = 1.472$), polymer
($n_{core} = 1.6$), silicon nitride ($n_{core} = 2.0$) and silicon ($n_{core} = 3.47$) waveguide

cores. The minimum extent of each curve is the point at which the fundamental waveguide mode is cut-off. Figure 9 clearly confirms that sensor response to adlayer adsorption increases with increasing core refractive index. The response also improves rapidly as the core thickness is decreased, but does reach a maximum below which the sensitivity drops. This maximum arises because the as the core is decreased further, waveguide confinement decreases and the mode spreads so far into the upper and lower cladding that relative overlap of the mode with the adsorbed layer (see equation 2) begins to decrease. The optimal thickness for a silicon waveguide sensor operating at this wavelength is 0.22 μm, which has an adlayer induced index change of approximately δN_{eff} = 0.003. Similar calculations were also carried out for TE polarized light, with similar results as shown in Figure 10. However the maximum induced δN_{eff} was about three times smaller than for the TM mode, as expected from the surface field enhancement originating from the electromagnetic boundary conditions leading to Eq. 4.

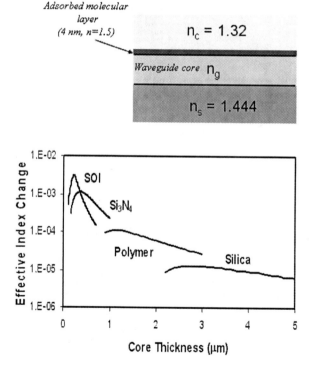

Figure 9. The effective index change induced by the adsorption of a 4 nm thick molecular layer with index n =1.5, calculated as a function of core thickness for the TM mode propagating in glass, polymer, silicon nitride and silicon waveguides.

Response simulations were repeated for the Si waveguide with a thin 10 nm layer of SiO$_2$ on the surface of the waveguide, as also shown in the inset of Figure 10. The new calculation shows that the sensor waveguide response to adsorption is only reduced by a few percent. Therefore Si sensor waveguides can be coated with a thin layer of SiO$_2$ with little loss in sensitivity, and the oxide surface functionalized to activate the sensor. This is a key result when considering the prospects for practical device implementation and commercialization, since most commercial micro-array devices are base on glass substrate chemistries. Therefore Si PWEF sensor can be functionalized for specific molecular targets, using well established chemistries borrowed from the biosensor micro-array industry.

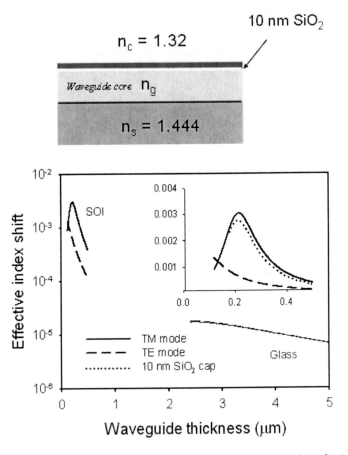

Figure 10. The calculated effective index change induced by the adsorption of a 4 nm thick molecular layer with index n =1.5, for the TE and TM modes in glass and silicon waveguides. The inset shows the results when a 10 nm thick SiO$_2$ layer is placed on the Si waveguide core.

5. Sensor Interrogation: Mach-Zehnder Interferometers and Ring Resonators

Molecular binding at the surface of the PWEF sensor is accomplished by monitoring the phase change of light propagating through the sensor waveguide. The simplest method for detecting such a phase change is through the use of the waveguide Mach-Zehnder interferometer shown in Figure 11. The incoming light is simply split into two equal parts by a 3 dB splitter and the two beams propagate independently along the two arms of the interferometer, and are recombined at a second 3 dB splitter. The total intensity in the output waveguide depends on the relative phase of the two beams when recombined at the final 3dB splitter.

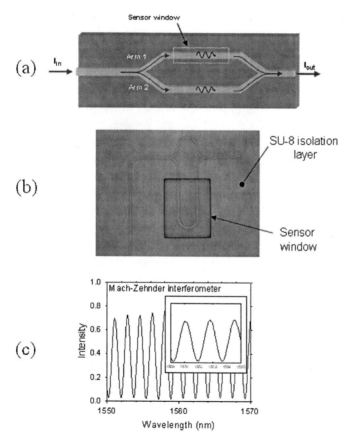

Figure 11. (a) A schematic diagram of an MZ inteferometer sensor circuit, and (b) a top view of a silicon waveguide MZ device as viewed from above using an IR camera to image the λ =1550 nm light scattered light. (c) The measured power from the output port as a function of wavelength for a silicon photonic wire MZ interferometer with unequal arm length.

The entire MZ interferometer is covered with protective layer as in Figure 8, and a window is opened over one of the MZ arms to expose a section of waveguide of length L to the sample fluid flowing above. As molecules bind in this sensor window the resulting phase change $\Delta\phi$ is given by Eq. 1, and the output intensity of the MZ interferometer will vary as

$$I = I_0\left[1-\cos(\Delta\phi)\right] = I_0\left[1-\cos\left(\delta N_{eff}\left(2\pi/\lambda\right)L\right)\right]. \qquad (5)$$

This equation assumes that both arms of the MZ interferometer are of equal length. If the interferometer is unbalanced, with arms of unequal length, the output intensity of the device will also vary with wavelength, as for the measured MZ output data shown in Figure 11(c). Nevertheless the output intensity at any fixed wavelength will still follow the form of Eq. 5 as δN_{eff} changes. An output signal of this form can easily be monitored in real time, and then converted to an absolute phase shift, and thereby to a measure of the number of molecules bound to the waveguide surface. Since the response is an oscillating function of index change, the intensity change with molecular adsorption will be highest halfway between maxima and minima of the output intensity, but zero at the maxima and minima. In practical use it may be useful optimize the response by biasing the MZ interferometer at the point of maximum slope. For given sensor dimensions, the detection limit is determined by the electrical and optical noise of the measurement system including light source noise, and noise in the photo-detector and controlling electronics. The overall detection and accuracy of the sensor output can also depend very strongly on environmental variables such as system temperature. For example, the index of refraction of silicon can vary by approximately 10^{-4} for a 1°C temperature change. Given the calculated response results given in Figure 9, the effective index change induced by heating even a few degrees can be comparable to the δN_{eff} arising from molecular adsorption. Fortunately, one of the advantages of the MZ layout is that by making the optical path length of the two arms precisely equal, common mode signals due to factors such as temperature which shift the effective index in both arms can be cancelled out.

The ring resonator (RR) shown in Figure 12 provides an alternative method for detecting a binding induced index change. Here light propagates along an input waveguide that is weakly coupled to a ring waveguide by a directional coupler[14] or a multimode interference coupler.[15]

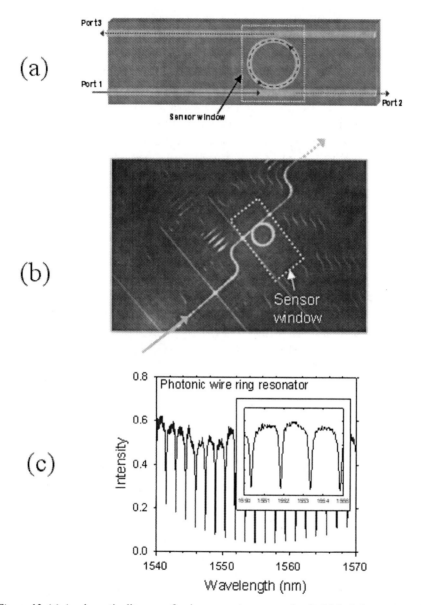

Figure 12. (a) A schematic diagram of a ring resonator sensor circuit. Light is input at port 1, and complementary output signals can be measured at port 2 or port 3. (b) A top view of a silicon waveguide ring resonator device as viewed from above using an IR camera to image the λ =1550 nm light scattered light. (c) The measured power from port 2 as a function of wavelength for a silicon photonic wire ring resonator.

When the wavelength is tuned so that an integer number m wavelengths is equal to the optical path $N_{eff}L$ around the ring, the light will be resonantly coupled into the ring waveguide. Here L is the total length of the ring. As a result there will be a decrease in intensity of light transmitted past the coupler and arriving at output port 2. An example of a measured output spectrum of a ring resonator is given in Figure 12 (c). If the ring is also coupled to a second output waveguide as in Figure 12(a), at resonance light will also couple strongly to this second waveguide and can be measured at the output port 3 in Figure 12. If a window is opened over the entire ring and molecules adsorb on the surface, the resonance wavelength will shift with the induced effective index change according to the relation

$$\delta\lambda_m = \delta N_{eff}\left(\frac{L}{m}\right) \qquad\qquad (6)$$

Either the wavelength shift, or the output intensity at a fixed wavelength within the resonance linewidth, can be an effective transduction signal for monitoring molecular binding. As with the MZ interferometer, the response is proportional to the length L of exposed waveguide. However, unlike the MZ based sensors, the line shape of the ring resonance can be very narrow if the quality factor (Q) of the ring is sufficiently high. By choosing the operating wavelength to be near the resonance peak, a small index shift can produce a large signal change at the output. On the other hand, the sensitivity to effective index shift is effectively zero between adjacent wavelength resonances, and there is no natural common mode rejection scheme for a sensor based on a single ring.

Although designing a sensor circuit with rings can therefore be more complicated, there are a number of ways of employing multiple rings to provide referencing or using active resonance tracking to resolve these issues. Eventually RR based sensor circuits may lead to much smaller individual sensor element that can be easily arrayed. For example in SOI, photonic wire waveguide ring resonators have been fabricated that have radii as small as 2 µm, however little work has been done to date beyond demonstration of single ring sensors[16–17].

The use of SOI photonic wire waveguides can offer obvious advantages from the point of view of designing the optical circuit. The bend loss of such waveguides is negligible for radii as small as a few micrometers. In comparison, in glass waveguides the bend radii cannot be much smaller than 5 mm. Hence the overall size of the entire optical sensor circuit can be orders of magnitude smaller using silicon that with other waveguide materials. This leads to the possibility of arraying a large number of sensor elements on a single chip, an essential step forward if waveguide based EF sensors are to compete with commercial micro-array biochips outside the research laboratory.

6. Silicon Photonic Wire Evanescent Field Sensors: Experiment

A number of MZ interferometer and RR based Si photonic wire based devices were fabricated to confirm the theoretical response predictions given in Figure 10, and demonstrate sensing in a simple affinity binding system. The device were all fabricated using SOI wafers with a 0.260 µm waveguide layer and a 2 µm thick buried oxide layer. The Si waveguides were formed by patterning the wafer using e-beam lithography, and then using inductively-coupled plasma etching to remove the silicon in the un-exposed areas completely down to the oxide layer. This produced rectangular Si channel waveguides with a width of 0.45 µm and height of 0.26 µm. As the waveguides approached the chip facets, the waveguides were adiabatically tapered down to approximately 0.1 µm to form inverse taper couplers.[18] This was done to improve coupling to the input fiber mode. The entire chip was then coated with 2 µm of SU-8 photoresist. Windows were opened in this SU-8 layer by optical contact lithography and etching, to expose the underlying waveguide in the active sensor area of the device. Finally a PDMS block with $100 \times 200 \ \mu m^2$ fluidic channels was aligned bonded to the chip surface so that sample fluid could be delivered to the sensor areas. Figure 13 shows a schematic diagram illustrating the sensor window and fluidic channel layout. The layout of each chip included a variety of ring and MZ interferometer based sensor circuits.

Experiments were carried out to measure the sensor response to both bulk liquid index changes and surface binding of Avidin protein to the surface. Light from a tunable laser, operating over the range from $\lambda = 1460$ nm to 1550 nm, was coupled into each device using a tapered fiber. Output light from the sensor circuit was collected using a microscope objective and measured using a photodetector.

The sensor response to homogeneous fluid index change was characterized by exposing the sensors to a series of sucrose solutions with incrementing refractive index. This was done for both unbalanced MZ interferometer and ring interrogation circuit designs. Using a syringe pump, the solutions were drawn from the reservoir through the microfluidic channel positioned over the sensing windows of the MZ and ring sensor elements. By monitoring the variation of the transmitted intensity spectrum with increasing index change, the induced phase shift of the light passing through the sensor waveguides was determined.

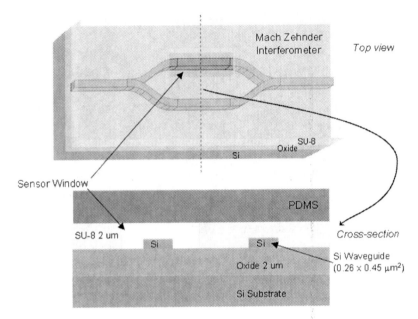

Figure 13. A schematic diagram of the sensor window in an MZ interferometer with the SU-8 isolation layer and PDMS fluidic channels.

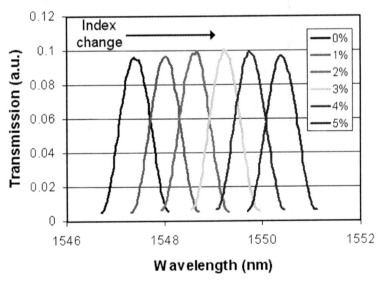

Figure 14. The wavelength shift of the output fringe pattern from an unbalanced MZ interferometer sensor with increasing fluid refractive index. For clarity only one period of the fringe pattern is shown. The exposed silicon photonic waveguide in one arm of the interferometer is 1.5 mm long.

Figure 14 shows the index induced shift of a single fringe from an unbalanced MZ interferometer with a $L = 1.5$ mm long active section in one arm of the device. Converting the measured wavelength shift to the phase shifts shown in Figure 15, an optical phase dependence of approximately $\Delta\phi / \Delta n_{clad} = 300 \times 2\pi$ per refractive index unit (RIU) was calculated. This agrees to within 15% of the phase shift sensitivity calculated using a two-dimensional, full-vectorial, mode expansion based waveguide solver. The corresponding measured index sensitivity is $\partial N_{eff} / \partial n_{clad} = 0.31$, at a signal wavelength of $\lambda = 1550$ nm. For comparison, the response of a free space beam propagating through the same medium would be only three times larger. Although our device does not have the optimum Si thickness, the measured index sensitivity $\partial N_{eff}/\partial n_{clad}$, shows approximately two orders of magnitude improvement over similar devices reported in silica waveguides,[19-20] and two to five times larger $\partial N_{eff} / \partial n_{clad}$ compared to more optimized Si_3N_4 waveguides.[21-23]

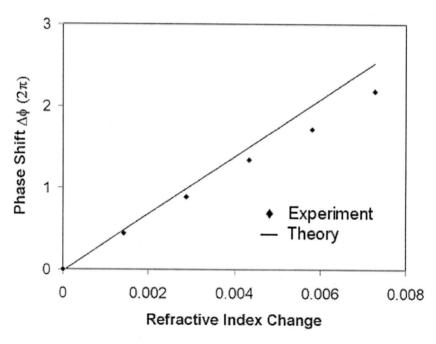

Figure 15. The variation in optical phase shift with fluid index calculated from the data in Figure 14 for a $L = 1.5$ mm long silicon PWEF sensor waveguide incorporated into a Mach-Zehnder interferometer interrogation circuit.

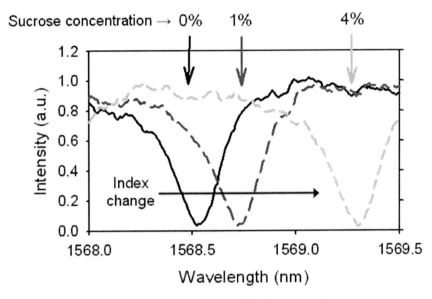

Figure 16. The shift in ring resonator line with fluid index change for a silicon PWEF in a ring resonator configuration with a 50 μm ring radius.

Figure 16 shows similar results for a ring resonator based sensor, using a ring with a 50 μm radius and overall sensor length of 314 μm. In this case the shift of the ring resonance is determined by monitoring the output intensity of port 2 in the schematic layout shown in Figure 12.

The sensor response to adsorption of a molecular monolayer was characterized using the biotin-avidin system. The biotin-avidin bond is one of the strongest molecular bonds found in nature. Hence this system is ideal for calibrating sensor response since the binding reaction proceeds in one direction only, with no desorption. Furthermore the avidin molecular weight and close-packed monolayer density are well known[24]. The sensor surface of a MZ interferometer sensor was functionalized by attaching a layer of biotin molecules to the surface of an $L = 1.5$ mm long exposed waveguide in the sensor window.

The output intensity of the MZ device was monitored as a PBS solution containing avidin molecules was drawn through the microfluidic channels and over the sensor waveguide surface. The output intensity as a function of time is shown in Figure 17(a). During the first 60 seconds after the onset of avidin flow, the output signal undergoes several rapid oscillations, after which the signal change slows dramatically. Figure 17 (b) shows the same data converted to optical phase using Eq. 5. This phase plot has the form of an adsorption isotherm, as expected since phase should be directly proportional to the density of adsorbed molecules. We interpret the rapid signal change in the first 60 seconds as the formation of a monolayer of avidin on the biotin coated surface. The ensuing much slower change in signal is attributed to non-specific adsorption (i.e. adsorption without binding to a biotin site) and rearrangements of the bound avidin-biotin complexes. Based on these curves and the observed signal to noise ratio, we expect that this sensor should be able to detect at as small as 1% of a monolayer of avidin molecules using the present experimental configuration. Assuming the molecular mass of avidin to be about 50 kDa, and the surface footprint of the molecule to be about 4 × 5 nm[24], we estimate that with no further optimization this sensor is capable of detecting well below 100 femtograms of a protein-like molecule.

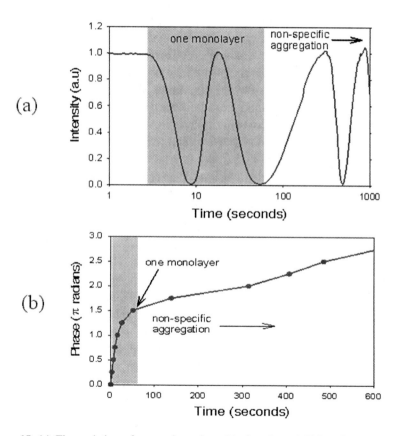

Figure 17. (a) The variation of output intensity with time for a MZ interferometer based sensor exposed to PBS solution of avidin molecules. The exposed sensor waveguide is 1.5 mm long. (b) The optical phase shift induced by avidin binding extracted form the data in (a). The shaded regions in both graphs indicate the regime during which a single monolayer is formed.

7. Conclusions

In this paper we have reviewed the potential of silicon photonic wire waveguides as extremely sensitive and compact evanescent field biosensors. Both simple physical arguments and numerical simulations are used to show that EF sensor response increases with increasing core refractive index and decreasing waveguide core size. This response enhancement arises because an increasing fraction of the waveguide modal power propagates in a narrow region extending about 100 nm above waveguide surface, precisely where molecular binding takes place. When the electric field polarization is normal to the waveguide surface, the surface electric field and hence sensor response can be further enhanced by the discontinuity of the normal electric field components. This effect is particularly strong in high index contrast waveguides such as silicon. The conclusion is that thin silicon photonic wire waveguides can offer the highest response to molecular adsorption of all commonly available waveguide systems. The Si sensor surface is covered by thin native SiO_2 (i.e. glass) film, but this film has a negligible effect on device response. More importantly, since the surface is glass, all established glass substrate functionalization and affinity chemistries used in the micro-array industry and in research may be directly adapted to the Si waveguide sensor. The effectiveness of the Si PWEF sensors has been demonstrated experimentally in both ring resonator and Mach-Zehnder interferometer configurations. The measured response of a Si waveguide sensor to bulk fluid index change is in good agreement with numerical simulations. Finally, the response to molecular adsorption has been characterized as well using the avidin-biotin system.

Evanescent field waveguide sensors provide highly sensitive label free detection of molecules using affinity binding interactions. These interactions may be based on DNA-DNA binding for genomic testing, or on antibody-antigen interactions for probing for the presence of a variety of proteins and organic molecules. Silicon sensor array element should outperform both other waveguide devices and SPR sensors with respect to sensitivity, but arrays of tens or hundreds of sensors can occupy only few square millimeters of silicon chip area, thus forming compact micro-arrays. By borrowing and scaling down multiplexing technologies from the telecommunications sector, each sensor array element can be monitored continuously so that full binding kinetics data from many sensors can be simultaneously collected in real time during exposure of the sensors to the sample. Since the substrates are silicon wafers, these chips are natural substrates for microfluidic elements and thermal or electrical actuators for controlling sample delivery and local temperature. In summary, silicon photonics appears to offers an unparalleled potential for combining integrated optics with micro-array biochip technologies to create a label free biosensiong and bioassay platform.

Acknowledgments

This work has been supported by the Genome and Health Initiative (GHI) at the National Research Council Canada.

References

1. R.G. Hunsperger, *Integrated Optics: Theory and Technology,* Third Edition, (Springer-Verlag, Berlin 1991).
2. G.T. Reed and A.P. Knights, *Silicon Photonics: An Introduction*, (John Wiley & Sons, Chichester, United Kingdom, 2004).
3. L. Pavesi and D.J. Lockwood, *Silicon Photonics,* (Springer-Verlag, Berlin, 2004).
4. H. Ouyang, C. Striemer, and P.M. Fauchet, Quantitative analysis of the sensitivity of porous silicon optical biosensors *Appl. Phys. Lett.* 88, 163108–163110 (2006).
5. J. Lu, C.M. Strohsahl B.L. Miller, and L.J. Rothberg, Reflective interferometric detection of label-free oligonucleotides, *Anal. Chem.* 76, 4416–4420 (2004)
6. J. Homola, Present and future of surface plasmon resonance biosensors, *Anal. Bioanal. Chem.* 377, 528–529 (2003).
7. J.J. Ramsden, Optical Biosensors, *J. Molecular Recognition* 10, 109–120 (1997).
8. B.J. Luff, R.D. Harris, J.S. Wilkinson, R. Wilson, and D.J. Schiffrin, Integrated optical detection biosensor, *J. Lightwave Technol.* 16, 583–592 (1998).
9. L.M. Lechuga, E. Mauriz, B. Sepulveda, J. Sanchez del Rio, A. Calle, G. Armelles, and C. Dominguez, Optical biosensors as early detectors of biological and chemical warfare agents, in *Frontiers in Planar Lightwave Technology*, S. Janz, J. Ctyroky, and S. Tanev, eds., 119–140 (Springer, Amsterdam, 2005).
10. H. Kogelnik, Theory of Optical Waveguides, in *Guided-Wave Optoelectronics*, 7-87, Second edition, 7–87, T. Tamir ed., (Springer-Verlag, Berlin 1990).
11. R.U. Ahmad, F. Pizutto, G.S. Camarda, R.L. Espinola, H. Rao, and R.M. Osgood, Ulracompact Corner-mirrors and T-branches in silicon-on-insulator," *IEEE Phot. Technol. Lett.* 14, 65–67 (2002).
12. V.R. Almeida, C.A. Barrios, R.R. Panepucci, M. Lipson, M.A. Foster, D.G. Ouzounov, and A.L. Gaeta, All-optical switching on a silicon chip, *Opt. Lett.* 29, 2867–2869 (2004).
13. T. Chu, H. Yamada, S. Ishida, and Y. Arakawa, Tunable add-drop multiplexer based on silicon-photonic wire waveguides", *IEEE Phot. Technol. Lett.* 18, 1409–1411 (2006).
14. W.R. Headley, G.T. Reed, S. Howe, A. Liu, and M. Paniccia, Polarization independent optical racetrack resonators using rib waveguides on silicon-on-insulator," *Appl. Phys. Lett.*, 85, 5523–5525, (2004).
15. D.-X. Xu, S. Janz, and P. Cheben, Design of polarization insensitive ring resonators in SOI using cladding stress engineering and MMI couplers, *IEEE Phot. Technol. Lett.* 18, 343–345 (2006).
16. A. Yalcin, K.C. Popat, J.C. Aldridge, T.A. Desai, J. Hryniewicz, N. Chbouki, B.E. Little, O. King, V. Van, S. Chu, D. Gill, M. Anthes-Washburn, M.S. Unlu and B.B. Goldberg,

Optical sensing of biomolecules using microring resonators, *IEEE J. Selected Topics in Quantum Electron.* 12, 148–154 (2006).

17. A. Densmore, D.-X. Xu, P. Waldron, S. Janz, A. Delâge, P. Cheben and J. Lapointe, Thin silicon waveguides for biological and chemical sensing, *Proceedings of the SPIE*, vol. 6477 (in press, 2007)

18. V.R. Almeida, R.R. Panepucci, and M. Lipson, Nanotaper for compact mode conversion, *Opt. Lett.* 28, 1302–1304, (2003).

19. B.J. Luff, R.D. Harris, J.S. Wilkinson, R. Wilson and D.J. Schiffrin, Integrated-optical directional coupler biosensor, *Opt. Lett.* 21, 618–619 (1996).

20. H. Ping, K. Kawaguchi, J.S. Wilkinson, Integrated optical dual Mach Zehnder interferometer sensor, in Proceedings of the Conference on Lasers and Electro-Optics, CLEO/Pacific Rim 2001, Vol. 1, No. 15-19, I-156-I-157 (2001).

21. E.F. Shipper, A.M. Brugman, C. Dominguez, L.M. Lechuga, R.P.H. Kooyman and J. Greve, The realization of an integrated Mach-Zehnder waveguide immunosensor in silicon technology, *Sensors and Actuators B* 40, 147–153, (1997).

22. F. Prieto, B. Sepulveda, A. Calle, A. Llobera, C. Dominguez, A. Abad, A. Montoya and L.M. Lechuga, "An integrated optical interferometric nanodevice based on silicon technology for biosensor applications," Nanotechnology, Vol. 14, pp. 907–912, 2003.

23. R.G. Heideman, R.P.H. Kooyman and J. Greve, "Performance of a highly sensitive optical waveguide Mach-Zehnder interferometer immunosensor," Sensors and Actuators B, Vol. 10, pp. 209–217, 1993.

24. J.M. Cooper, J. Shen, F.M. Young P. Connolly, J.R. Barker and G. Moores, "The imaging of streptavidin and avidin using scanning tunneling microscopy" J. Materials Science: Materials in Electronics 5, pp. 106–110 (1994) avidin properties

SELECTIVITY OF HYBRIDIZATION CONTROLLED
BY THE DENSITY OF SOLID PHASE SYNTHESIZED DNA
PROBES ON GLASS SUBSTRATES

FAYI SONG AND ULRICH J. KRULL*
*Department of Chemical and Physical Sciences, University
of Toronto, 3359 Mississauga Road, Mississauga, Ontario,
Canada, L5L 1C6*

Abstract. Optical biochip design based on varying the density of immobilized single-stranded DNA (ssDNA) oligonucleotide probes was examined. A method of immobilization was developed to yield various densities of probe molecules using photochemical activation of surfaces and *in situ* solid phase synthesis for DNA immobilization. High surface density of ssDNA probe (up to 1×10^{13} probes/cm^2) was obtained using the immobilization method. The densities and extent of hybridization of nucleic acids were determined using confocal fluorescence microscopy. Selective hybridization of targets associated with spinal muscular atrophy containing single nucleotide polymorphisms (SNP), and their thermal denaturation profiles were investigated to examine the sensitivity and selectivity for SNP detection. The detection limit was less than 16 pM at room temperature. Single base mismatch discrimination was demonstrated based on control of melt temperature by selection of probe density, and temperature differences of 12–15°C could be achieved for SNP determination. Importantly, the results demonstrate that poor control of probe density can result in significant variability of selectivity, as seen by melt temperature shifts of up to 5°C in the density range that was investigated.

Keywords: Nucleic acid; hybridization; fluorescence; density; single nucleotide polymorphism; immobilization.

---

*To whom correspondence should be addressed. U. J. Krull, e-mail: ukrull@utm.utoronto.ca.

W.J. Bock et al. (eds.), Optical Waveguide Sensing and Imaging, 195–210.
© 2008 *Springer.*

1. Introduction

Most DNA microarrays and biosensors provide selectivity based on the hybridization of target DNA to immobilized single-stranded DNA (ssDNA) probes. Sensitivity and selectivity are typically the key factors considered in the development of nucleic acid biosensors and microarrays (Piunno and Krull 2005). Selectivity is very important for applications in areas such as clinical diagnostics where single nucleotide polymorphism (SNP) detection is of significant interest (Oh et al., 2005; Watterson et al., 2004). Thermodynamic considerations are often evaluated as an approach to manipulate selectivity. It is clear from this perspective that selectivity is not just a function of the nucleic acid sequence that is used to define an immobilized probe molecule. The thermodynamic stability of double-stranded DNA (dsDNA) is dependent on nearest-neighbour and surface interactions, including the extent of surface occupancy. This has consequences in terms of selectivity, the efficiency of binding, and kinetics, where each can change as a result of the local environment and the extent of formation of hybrids during an analytical experiment (Dodge et al., 2004; Fortin et al., 2005; Peterson et al., 2000; Piunno et al., 2005; Shchepinov et al., 1997; Steel et al., 2000; Watterson et al., 2002; Watterson et al., 2000; Watterson et al., 2004; Zeng et al., 2003).

Probe immobilization density is a fundamental consideration for microarray and biosensor technologies in terms of sensitivity and selectivity. Probe density can be controlled by varying pre-synthesized ssDNA strand concentration and reaction time for immobilization onto gold or other functionalized substrates including glass slides (Jin et al., 2003; Levicky et al., 1998; Levicky and Horgan 2005; Peterson et al., 2000; Peterson et al., 2001; Steel et al., 2000). The resulting density of surface immobilized probes is somewhat limited due to surface occlusion and electrostatic repulsion between negatively charged ssDNA strands that become concentrated at the surface. Alternatively, probe density can be controlled by adjusting reaction conditions when using in situ synthesis of oligonucleotides (Beier and Hoheisel, 1999; Cattaruzza et al., 2006; Egeland and Southern, 2005; Gao et al., 2004; Gao et al., 2001; Pirrung et al., 2001; Watterson et al., 2000).

Among a variety of methods that are used for in situ synthesis of oligonucleotides, light-directed synthesis of DNA arrays on glass substrates provides an efficient and versatile method for high-density DNA chip fabrication (Fodor et al., 1991; Pease et al., 1994). Many photolabile groups, such as ((-methyl-2-nitropiperonyl)-oxy)carbonyl (MeNPOC) (McGall et al., 1997), dimethoxybenzoincarbonate (DMBOC) (Pirrung et al., 1998; Piunno et al., 2005), 2-(2-nitrophenyl)-propoxycarbonyl (NPPOC) (Beier and

Hoheisel, 1999; Beier and Hoheisel 2000; Hansan et al., 1997; Beier et al., 2001; Pirrung et al., 2001), have been used for photochemical DNA synthesis.

There have been very few reports about systematic studies of the effects of the density of immobilized probe molecules on the sensitivity and selectivity of DNA detection using hybridization on microarrays (Piunno and Krull, 2005) and as continuous films on substrates (Park and Krull, 2006). Several techniques, such as microfluidic devices (Jiang et al., 2005) and electrochemical control (Plummer et al., 2003; Terrill et al., 2000), have been reported for generation of immobilized gradients of density of probe molecules (Chaudhury and Whitesides, 1992; Dertinger et al., 2002; Mougin et al., 2005). The work reported by Park (Park and Krull, 2006) was limited in terms of the density of immobilization that was achieved, and also by poor reproducibility in the quality within and between immobilized films of probe molecules. The new work herein was intended to develop an improved method of immobilization that could provide for higher probe densities and improved film quality, so that it would be possible to quantitatively determine the range of selectivity that could be achieved for a single type of probe molecule by adjustment of immobilization density. An approach using a photochemical activation of surfaces has been developed in combination with in situ solid phase synthesis to prepare pads of ssDNA of various densities with good reproducibility, and these have been examined to determine effects of density on selectivity and sensitivity by use of fluorescence imaging.

2. Experimental

2.1. MATERIALS

Reagents for DNA synthesis, tetrazole, Cy5-phosphoramidite, spacer phosphoramidite 18 (18-O-Dimethoxytritylhexaethyleneglycol,1-[(2-cyano-ethyl)-(N,N-diisopropyl)]-phosphoramidite), and UniCap phosphor-amidite (Diethyleneglycol ethyl ether (2-cyanoethyl)-(N,N-diisopropyl)-phospho-ramidite) were from Glen Research Corp., Virginia, USA. Anhydrous acetonitrile, pyridine, toluene, hexaethylene glycol (HEG), dimethoxytrityl (DMT) chloride, diisopropylethylamine, ethylenediamine, and 3-glycido-xypropyltrimethoxysilane (GOPS) were from Sigma-Aldrich and were stored in glove box to maintain anhydrous conditions. 2-(2-nitrophenyl)-propoxycarbonyl (NPPOC)-deoxynucleotide (dT) phosphor-amidites from Proligo LLC (CO, USA) were directly used without further purification. N-(3-triethoxysilylpropyl)-4-hydroxybutyra-mide was from Gelest Inc., PA, USA. Water was double-distilled in glass and autoclaved. Silica gel

(Toronto Research Chemicals, Toronto, ON) had a particle size of 30–70 μm. Glass slides (76 × 25 cm) were from Fisher Scientific. The oligonucleotide sequences used as immobilized probe and labeled targets in thermal denaturation are shown below. Immobilized SMN1 probe: 5'-A TTT TGT CT*G* AAA CCC TGT-3', complementary Cy3-labeled target SMN1A: 5'-Cy3-ACA GGG TTT C*AG ACA AAA T-3', and single base mismatched Cy3-labeled target SMN2A: 5'-Cy3-ACA GGG TTT *T*AG ACA AAA T-3'. Presynthesized DNA targets were from Integrated DNA Technologies, Inc., IA, USA. All buffers were made using chemicals from Sigma-Aldrich Canada (Oakville, ON, Canada) and were prepared with autoclaved double-distilled water. Hybridization/washing solutions of 6x, 3x and 1xSSPE buffer with 0.1% (w/v) polyvinylpyrro-lidone (PVP; average molecular weight 36,000 daltons) were made by diluting 20xSSPE stock solution (1.5 M NaCl, 200 mM NaH_2PO_4, 20 mM EDTA pH 7.5) and dissolving PVP in the solution.

2.2. PROCEDURES

2.2.1. *Cleaning and Functionalization of Glass Slides*

The glass slide substrates were added to a 1:1:5 (v/v/v) solution of 30% ammonia/30% H_2O_2/deionized water at 80°C for 5 min. The substrates were then recovered, washed with water and then added to a 1:1:5 (v/v/v) conc. HCl/30% H_2O_2/deionized water at 80°C for 5 min. The substrates were then washed with water (3 × 250 mL), methanol (3 × 250 mL), chloroform (2 × 250 mL), and diethyl ether (2× 250 mL), respectively, dried over anhydrous $CaSO_4$ under vacuum until required.

The cleaned glass slides were soaked in a solution of toluene/ GOPS/ diisopropyl ethylamine (100:30:1 (v/v/v)) and the mixture was stirred at 85°C for 24 h under Ar. The substrates were then washed with methanol (2 × 250 mL), CH_3Cl (2 × 250 mL), Et_2O (2 × 250 mL) successively, dried and stored under vacuum and anhydrous $CaSO_4$ at room temperature.

2.2.2. *Synthesis and Linkage of DMT-HEG*

Synthesis of 17-dimethoxytrityloxa-3,6,9,12,15-pentaoxa-1-hepta-decanol (DMT-HEG) used 3.6 g of dimethoxytrityl chloride in 10 mL absolute pyridine and was added drop-wise to 3.0 mL of hexaethylene glycol in 5 mL absolute pyridine under Ar. The mixture was stirred overnight. After TLC analysis (hexanes:ethyl acetate 1:1) to ensure the complete absence of dimethoxytrityl chloride, the mixture was diluted using 300 mL of CH_2Cl_2, and was then washed by 5% $NaHCO_3$ (2 × 50 mL), water (2 × 50 mL) and brine (100 mL). The solvent was removed and the residue was purified by silica gel column chromatography to yield 2.1 g DMT-HEG (35% yield). 1H-NMR (200 MHz, CDCl$_3$) δ = 3.22 (t, 2H), 3.51–3.74 (m, 22H), 3.78 (s, 6H), 6.81 (d, 4H), 7.19–7.47(m, 9H).

Linkage of DMT-HEG onto GOPs functionalized glass slides used 1 g (1.7 mmol) of DMT-HEG dissolved into 10 mL, which was added to 800 mg (20 mmol) of NaH suspended in 10 mL anhydrous pyridine. The mixture was stirred at room temperature for 2 h under Ar, followed by filtration through a sintered glass frit under a positive pressure of Ar into a round bottom flask containing GOPS functionalized glass slides. The mixture was stirred at room temperature for 8 h under Ar. Then the glass slides were washed with methanol (3 × 250 mL), water (3 × 250 mL), methanol (3 × 250 mL), CH$_2$Cl$_2$ (3 × 250 mL), Et$_2$O (3 × 250 mL), respectively. The DMT-protected HEG functionalized glass slides were dried and stored under vacuum and over anhydrous CaSO$_4$.

A different process using N-(3-triethoxysilylpropyl)-4-hydro-xybu-tyramine instead of GOPS for the silanization of the surface provided a means to avoid use of NaH. Clean glass slides were soaked in 1% N-(3-triethoxysilylpropyl)-4-hydroxybutyramide in 95% ethanol overnight. After thoroughly rinsing with ethanol and deionized water, the glass slides were heated at 110°C for 20 min. The DMT-protected HEG spacer could be then immobilized onto the free hydroxyl groups at the surface. This was done by using HEG phosphoramidite 18 in acetonitrile in conjunction with the standard phosphoramidite method. In terms of density and functionality the results were similar to those obtained when using GOPS. For consistency, all data reported herein are based on use of GOPS in the immobilization procedure.

2.2.3. Immobilization of Photolabile Groups and Photodeprotection

The immobilization of photolabile groups and photodeprotection were done using a customized flow injection reactor cell. The flow cell was mounted on an optical bench with a XYZ stage (Thorlabs, Inc., NJ), and the patterning was controlled manually by moving the stage. Delivery of reagents to the reactor cell was done by pumping carrier solvent (acetonitrile) through an injection valve (Valco Instruments Co. Inc., TX). The DMT protecting groups were removed from the glass slides that were covered with DMT-HEG layers by treatment with 2% dichloromethane. The photolabile groups (NPPOC) were then immobilized onto the surface by phosphoramidite coupling using 75 mM NPPOC-dT phosphoramidite in acetonitrile for 5 min, where 0.5 M tetrazole in anhydrous acetonitrile was used as the activator. Any remaining free hydroxyl groups were then capped by coupling with 75 mM UniCap phosphoramidite. Oxidations were done by injection of 250 mL 0.05 M iodine in acetonitrile/pyridine/water (7:1:2, v/v/v).

Photodeprotection was done using a General Electric 85W H85A3 UV mercury arc lamp. To remove the NPPOC photolabile groups, the carrier solvent was changed to 0.05 M piperidine in acetonitrile, which was pumped through the flow cell during photodeprotection. Irradiation was done through apertures that were 1 mm in diameter. Various irradiation times varying from 1 min to 90 min were applied to different spots in order to obtain pads containing different densities of ssDNA.

2.2.4. *Solid Phase Phosphoramidite Synthesis of Oligonucleotides*

DNA synthesis was done on the functionalized glass slides after photo-deprotection was completed. An PE-ABI 392-EP DNA synthesizer (Perkin-Elmer Applied Biosystems, Foster City, CA) was combined with a custom-mized flow cell to replace the conventional reaction column. Detritylation was done using 2% dichloroacetic acid (DCA) in dichloromethane (DCM). Activation was done using 0.5 M tetrazole in anhydrous acetonitrile and the coupling time used was 150 seconds. Oxidation was done with 0.05 M iodine in acetonitrile/pyridine/water (7:1:2, v/v/v) for 1 min. After synthesis of the 20mer DNA probes, the glass slide with the flow cell was transferred to the flow injection system for labelling with Cy5 phosphoramidite.

2.2.5. *Fluorescence Labelling and Imaging*

The free hydroxyl groups produced on the surfaces of slides after photo-deprotection of the photolabile groups and/or removal of DMT groups at the end of DNA synthesis were labeled by coupling with Cy5 phosphoramidite (5 mM in a solution containing 45 mM DMT-T phosphoramidite) for 3 min, followed by 3 min oxidation with 0.05 M iodine solution. Then the Cy5 monomethoxytrityl (MMT) protecting groups were removed by manual injection of 2% DCA in DCM. Finally, the slides were removed from the flow cell and soaked in a solution of 50% ethylenediamine in ethanol for 2 hours at room temperature to remove the base- and phosphate-protecting groups. Before fluorescence scanning was done, the slides were rinsed thoroughly with ethanol, acetone/methanol/water (1:1:1), 6xSSPE 0.1% PVP, 3xSSPE 0.1% SDS, 3xSSPE buffer, and deionized water, respectively.

All fluorescence data and images were collected using a VersArray ChipReader version 3.1 (Bio-Rad Laboratories, Inc, CA) with both a 532 nm diode laser (Cy3 channel) and 635 nm diode laser (Cy5 channel) for excitation. The output was in the form of 16-bit TIFF files. ImageQuant 5.2 (Molecular Dynamics) was used for data analysis of the scanned images. A resolution 25 μm was used for collection of fluorescence signal intensity. The average intensity of pixels within every spot was used for quanti-fication. Every spot contained approximately 550 pixels.

2.3. HYBRIDIZATION AND REGENERATION OF THE IMMOBILIZED PROBES

Freshly prepared slides coated with immobilized single-stranded DNA probes (ssDNA) were activated by soaking in 85°C 6xSSPE 0.1% PVP hybridization solution containing 80 nM fully complementary target DNA for 5 min and then were allowed to cool to room temperature (22±2°C). This was repeated twice and then the slides were washed, i.e., hybridization solution, 3xSSPE 0.1% SDS solution, and finally 3xSSPE solution for 5 min, respectively. After preliminary surface fluorescence analysis, the immobilized ssDNA probes were regenerated by removing the target DNA for the next hybridization experiments. Typical hybridization experiments were done by placing 50 μL of Cy3-labeled target DNA solution on the slide, followed by storage of the slide in a high-humidity chamber to avoid evaporation of water for a period of one hour. Non-specifically adsorbed and excess material was eliminated with the washing procedure. The slide was dried using an argon stream and then fluorescence scanning measurements were completed. Regeneration was done by immersing the slide into 85°C stripping solution (2.5 mM Na_2HPO_4, 0.1% SDS) and deionized water to remove the hybridized target DNA from the surface. The slide coated with ssDNA probes was stored in a vacuum desiccator or in a solution of ethanol-water (1:1) to avoid contamination during storage.

2.4. ACQUISITION OF SURFACE THERMAL DENATURATION PROFILES

Thermal denaturation profiles from the glass slides were acquired by monitoring fluorescence from Cy3-labeled target DNA (SMN1A) over the temperature range of 25–80°C with temperature increments of 5°C and 5 min equilibrium at each temperature. Different ionic strength buffer solutions including 6xSSPE, 3xSSPE and 1xSSPE with 0.1% PVP were used to investigate the effect of ionic strength on the thermal denaturation profiles. Slides were removed from the solutions and dried with argon for the fluorescence scanning analysis.

3. Results and Discussion

3.1. IMMOBILIZATION OF PHOTOLABILE GROUPS AND PHOTODEPROTECTION

There are several kinds of photosensitive protecting groups that have been developed for photolithographic DNA synthesis. The 2-(2-nitrophenyl)-propoxycarbonyl (NPPOC) group has been identified as particularly promising.

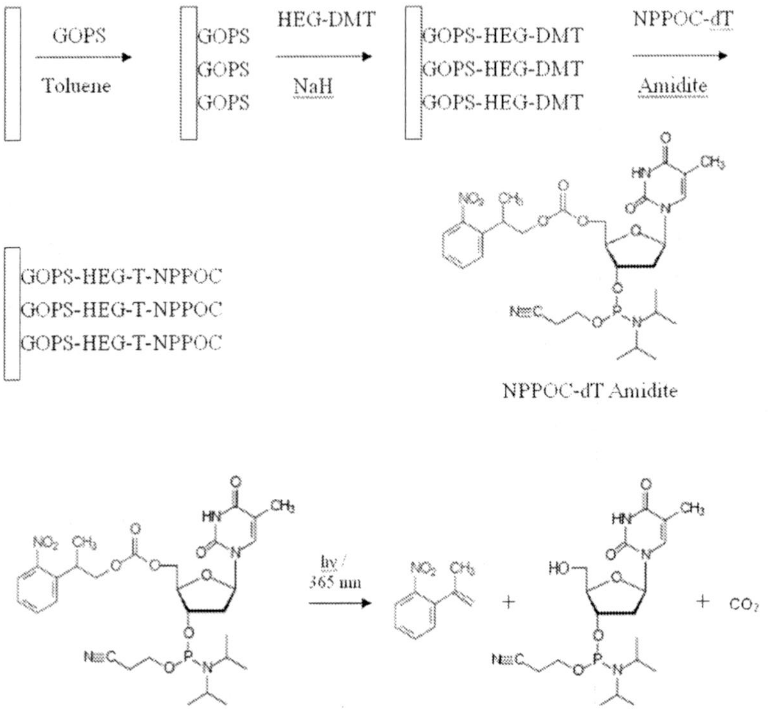

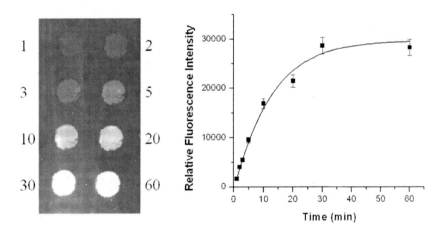

Scheme 1. Photochemical deprotection for activation of surface and growth of DNA by automated synthesis.

Figure 1. Fluorescence image of a glass wafer after removal of NPPOC groups and staining with Cy5 phosphoramidite. Fluorescence intensity was dependent on time of UV irradiation.

NPPOC groups can be efficiently deprotected photochemically when the process is done in the presence of appropriate amounts of amine base (Beier and Hoheisel 1999, 2000), and the reaction is catalyzed by UV radiation rather than room light so that control of irradiation is improved. Therefore, NPPOC-dT phosphoramidite was chosen and base-catalyzed conditions were adopted in this work. Scheme 1 shows the process for the immobilization of NPPOC groups on glass slides. After cleaning and silanization, DMT protected HEG was immobilized onto the glass slides as a spacer. Following removal of DMT from the surface by injection of 2% DCA in DCM, NPPOC-dT phosphoramidite was injected to the flow cell to react with the free hydroxyl groups so that NPPOC-dT was immobilized on the slide using the standard tetrazole-catalyzed phosphoramidite coupling. Any remaining free hydroxyl groups on the surface were capped with UniCap phosphoramidite. After oxidation, the slides were ready for photodeprotection.

Photochemical activation of the surface and removal of NPPOC groups from the slides was done with irradiation 365 nm. According to the reported photochemical reaction mechanism (Beier and Hoheisel 2000; Hansan et al., 1997; Pirrung et al., 2001), the NPPOC groups at the 5' position of nucleoside phosphoramidite can be removed and free hydroxyl groups are produced as shown in Scheme 1. Irradiation was done for various lengths of time to control the density of hydroxyl groups.

The density of hydroxyl groups on the glass was determined by reaction with Cy5 phosphoramidite and subsequent measurement of fluorescence intensity, and this method was used to follow the surface photolysis rates. To ameliorate concerns about potential fluorescence quenching effects between neighboring fluorophores on the surface, a solution of 5 mM Cy5 phosphoramidite and 45 mM DMT-dT phosphoramidite in acetonitrile was used for labeling experiments (Beier and Hoheisel 2000; McGall et al., 1997). Figure 1 provides an example of fluorescence intensity for various irradiation times and indicates saturation between about 60-90 min of irradiation. The relative standard deviation of intensity distribution for each of the spots was about 5%.

3.2. SOLID PHASE SYNTHESIS AND IMMOBILIZATION OF DNA PROBES

The standard phosphoramidite protocols for *in situ* solid phase synthesis of DNA sequences were used to prepare high densities of ssDNA that were conducive to improvement of selectivity for SNP detection. The hydroxyl groups created by removal of NPPOC provided the starting point for the subsequent DNA synthesis. The resulting density of surface synthesized DNA probes was templated by the density of the hydroxyl groups. The probe DNA sequence and length could be readily controlled using the DNA synthesizer parameters. The sequences dT20 and SMN1 were synthesized onto the functionalized glass slides.

After completion of the DNA synthesis, Cy5 phosphoramidite was incorporated at the end of each DNA strand so that the immobilized DNA probes could be visualized by fluorescence intensity measurements. The fluorescence intensity increased with increased irradiation time. The results were consistent with those observed for direct staining of free hydroxyl groups with Cy5 phosphoramidite after photochemical removal of NPPOC protecting groups. This indicated that the immobilization of oligo-nucleotides occurred primarily at the sites produced by photo-deprotection, and the density of ssDNA could be controlled by the irradiation time.

Surface densities of the dT20 strands synthesized on glass slides were estimated by comparison with fluorescence intensities for known quantities of Cy5 labeled presynthesized dT20 strands that were spotted on the same slides. The highest density of in $situ$ synthesized dT20 strands was found to be 1×10^{13} strands/cm^2. When considering the length of the dT20-HEG conjugate (ca. 100 Å) (Watterson et al., 2000), this packing density is sufficiently high that there will be interactions between neighbouring strands that can influence the duplex stability (Peterson et al., 2001; Watterson et al., 2000).

Oligonucleotide probes that were 19 nucleotides in length and selective for the SMN1 gene fragment, which is associated with spinal muscular atrophy (SMA), were synthesized on glass slides. As shown in Figure 2A, the density of SMN1 probes could be controlled on glass slides. The trends were similar to those seen for dT20 immobilization, indicating a good reliability and reproducibility of the immobilization process.

3.3. SELECTIVITY OF DNA HYBRIDIZATION BASED ON DENSITY OF DNA PROBES

Hybridization of Cy3 labeled SMN1A target DNA with the SMN1 probe DNA was done. Figure 2B shows data from the Cy3 fluorescence channel and confirms that hybridization reflects the density of probe molecules. This is different from the results reported by Shchepinov et al. (Shchepinov et al., 1997), where the yield of hybridization at 36°C for 100% surface coverage of in $situ$ synthesized DNA probes using the phosphoramidite method was only one-half of that for a 50% surface coverage of DNA probes. Peterson et al. studied the effect of surface probe density on DNA hybridization using SPR (Peterson et al., 2000; Peterson et al., 2001). They found that the hybridization efficiency was 100% and the kinetics of binding

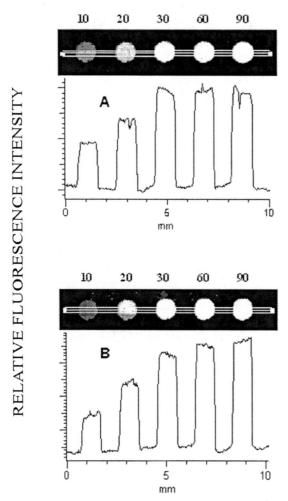

RELATIVE FLUORESCENCE INTENSITY

Figure 2. Fluorescence images and the cross-section fluorescence intensity profiles of (A) *in situ* synthesised SMN1 probes (Cy5 labelled); (B) Cy3 labeled complementary target SMN1A hybridized to SMN1 probes. The numbers over the image show the irradiation time (min). Hybridization time was 60 min, using a 1.6 μM SMN1A solution at room temperature.

followed comparatively faster kinetics in low probe density regimes. The total yield of hybridization did not decrease with increasing probe density in their results, and is in agreement with our results.

Figure 3 illustrates the dependence of surface fluorescence intensity for SMN1A concentration. The fluorescence intensity recorded for the Cy3 labeled SMN1A hybridized on the 5 pads of different densities of SMN1 probes increased with increasing SMN1A concentration from 16 pM to 160 pM. Linear response was observed for all the five density spots in this concentration range. The correlation coefficients of the linear fits using the method of least squares are better than 0.99. Overall, the results illustrate that a high surface density of oligonucleotide probes is suitable for high sensitivity of detection of DNA hybridization.

Single base mismatch discrimination is clearly demonstrated by the significant difference (12–15°C) in melt temperature observed from the profiles. The highest T_m value is 48°C for the fully matched hybrids. Interestingly, both the melt temperature T_m and ΔT_m for both the fully matched and single base mismatched hybrids increased with increasing surface density of SMN1 probes, and the melting transition curve became somewhat sharper. This is different from previous results obtained using fiber-optic biosensors in a total internal reflection fluorescence instrument (Watterson et al., 2000; Watterson et al., 2004), in which melt temperatures decreased with increasing surface density of DNA probes with 10^{-7} M target DNA present in buffer solution during data collection at different temperatures. One important aspect to consider is that complete deprotection of immobilized probes in films prepared by automated synthesis is a significant challenge. It is possible using this UV catalyzed synthesis that the extent of deprotection decreased with increasing density of a film.

Recovery of synthesized sequences from glass surfaces that were fabricated using this method suggested that synthesis that was not as efficient (90%) as reported earlier (95%) using wet chemical deprotection (Watterson et al., 2000; Watterson et al., 2004). Ultimately, it was clear that thermodynamic stability of hybrids as judged by melt temperatures could be reproducibly associated with density when using the NPPOC deprotection scheme in automated synthesis.

The experimental data shown in Figure 4 indicate that selectivity can be controlled by adjustment of the density of ssDNA probe, and also suggest the range of DNA density that is suitable to provide selectivity control for sensing systems that are based on gradients of probe density built by use of the NPPOC deprotection scheme (Park and Krull, 2006). A consistent effect of surface probe density on denaturation of dsDNA at room temperature was also noted where the hybrids at low surface density were more readily removed from the surface by washing. This finding is perhaps more significant in a practical context, as denaturation provides better selectivity for biosensor development than does the extent of hybridization.

Surface density ■<●<▲<▼<★

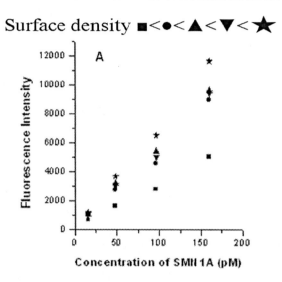

Figure 3. Concentration dependence of the surface fluorescence intensity for the SMN1A target hybridized on SMN1 probes. The order of the surface density of SMN1 probes (see symbols over Fig. 3) was approx. 0.28, 0.59, 0.79, 0.93, 1.02 × 10^{13} probes/cm^2. Hybridization time was 90 min in 6xSSPE 0.1% PVP solution.

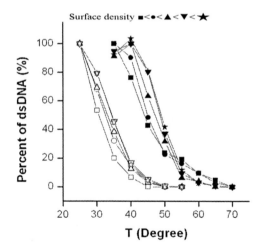

Figure 4. Thermal denaturation profiles for the fully matched SMN1A (solid symbols) and single base-pair mismatched SMN2A (open symbols) hybrids using SMN1 probes at different surface densities. The order of the surface density of SMN1 probes (see symbols over Fig. 4) was approx. 0.28, 0.59, 0.79, 0.93, 1.02 × 10^{13} probes/cm^2. 1xSSPE 0.1% PVP buffer solution was used for the thermal denaturation at different temperatures. Hybridization time was 60 min in a 6xSSPE 0.1% PVP solution containing 1.6 μM SMN1A or SMN2A at room temperature.

4. Conclusions

A method for controlling density of oligonucleotide probes immobilized on glass slides in an array format was demonstrated, and has been achieved by photochemical activation of surface followed by *in situ* solid phase DNA synthesis. The method was able to produce high densities of immobilized ssDNA on glass. At a high density of SMN1 probes at room temperature, hybridization response to single mismatched target DNA showed only 30-40% of the response to the fully complementary DNA. Thermal control was investigated to improve SNP detection selectivity. Experimental results indicated that the melt temperature increased with increasing surface density of the immobilized DNA. A large difference (up to 15°C) in melting temperature between fully complementary target DNA and single base-pair mismatched DNA hybridized with SMN1 probes was obtained. The results indicated that low surface density facilitated destabilization of the surface immobilized hybrids. Importantly, the melt temperature results demonstrate a range of about 5°C between films of densely immobilized probe and films that are of lower density (1 to 0.3×10^{13} probes/cm^2). This signals that poorly controlled density control as might be seen in spotted microarrays would result in significant variation of selectivity for identical immobilized nucleic acid probes.

Acknowledgements

We are grateful to the Natural Sciences and Engineering Research Council of Canada for financial support of this work. Special thanks are given to Drs Paul A.E. Piunno, Xuezhu Liu, and Xiaofeng Wang for useful discussion.

References

Beier M., and Hoheisel J. D., 1999, Versatile derivatisation of solid support media for covalent bonding on DNA-microchips, *Nucleic Acids Res.* 27 (9): 1970–1977.

Beier M., and Hoheisel J. D., 2000, Production by quantitative photolithographic synthesis of individually quality checked DNA microarrays, *Nucleic Acids Res.* 28 (4): 11.

Beier M., Stephan A., and Hoheisel J. D., 2001, Synthesis of Photolabile 5'-O-Phosphoramidites for the Photolithographic Production of Microarrays of Inversely Oriented Oligonucleotides, *Helv. Chim. Acta* 84: 2089–2095.

Cattaruzza F., Cricenti A., Flamini A., Girasole M., Longo G., Prosperi T., Andreano G., Cellai L., and Chirivino E., 2006, Controlled loading of oligodeoxyribonucleotide monolayers onto unoxidized crystalline silicon; fluorescence-based determination of the surface coverage and of the hybridization efficiency; parallel imaging of the process by Atomic Force Microscopy, *Nucleic Acids Res.* 34 (4): e32.

Chaudhury M. K., and Whitesides G. M., 1992, Correlation between surface free-energy and surface constitution, *Science* **255**: 1230–1232.

Dertinger S. K. W., Jiang X., Li Z., Murphy V. N., andWhitesides G. M., 2002, Gradients of substrate-bound laminin orient axonal specification of neurons, *Proc. Natl. Acad. Sci. USA* **99** (20): 12542–12547.

Dodge A., Turcatti G., Lawrence I., de Rooji N. F., and Verpoorte E., 2004, A Microfluidic Platform Using Molecular Beacon-Based Temperature Calibration for Thermal Dehybridization of Surface-Bound DNA, *Anal. Chem.* **76** (6): 1778–1787.

Egeland R. D., and Southern E. M., 2005, Electrochemically directed synthesis of oligonucleotides for DNA microarray fabrication, *Nucleic Acids Res.* **33** (14): e125.

Fodor S. P. A., Read J. L., Pirrung M. C., Stryer L., Lu A. T., and Solas D., 1991, Light directed, spatially addressable parallel chemical synthesis, *Science* **251**: 767–773.

Fortin E., Defontaine Y., Mailley P., Livache T., and Szunerits S., 2005, Micro-Imprinting of Oligonucleotides and Oligonucleotide Gradients on Gold Surfaces: A New Approach Based on the Combination of Scanning Electrochemical Microscopy and Surface Plasmon Resonance Imaging (SECM/ SPR-i), *Electroanalysis* **17** (5-6): 495–503.

Gao X., Gulari E., and Zhou X., 2004, *In situ* synthesis of oligonucleotide microarrays, *Biopolymers* **73**: 579–596.

Gao X., LeProust E., Zhang H., Srivannavit O., Gulari E., Yu P., Nishiguchi C., Xiang Q., and Zhou X., 2001, A flexible light-directed DNAchip synthesis gated by deprotection using solution photogenerated acids, *Nucleic Acids Res.* **29** (22): 4744–4750.

Hasan A., Stengele K., Giegrich H., Cornwell P., Isham K. R., Sachleben R. A., Pfleiderer W., and Foote R., 1997, Photolabile protecting groups for nucleosides: Synthesis and photodeprotection rates, *Tetrahedron* **53** (12): 4247–4264.

Jiang X., Xu Q., Dertinger S. K. W., Stroock A. D., Fu T., and Whitesides G. M., 2005, A General Method for Patterning Gradients of Biomolecules on Surfaces Using Microfluidic Networks, *Anal. Chem.* **77** (8): 2338–2347.

Jin R., Wu G., Li Z., Mirkin C. A., and Schatz G. C., 2003, What controls the melting properties of DNA-linked gold nanoparticle assemblies?, *J. Am. Chem. Soc.* **125** (6): 1643–1654.

Levicky R., Herne T. M., Tarlov M. J., and Satija S. K., 1998, Using self-assembly to control the structure of DNA monolayers on gold: A neutron reflectivity study, *J. Am. Chem. Soc.* **120**: 9787–9792.

Levicky R., and Horgan A., 2005, Physicochemical perspectives on DNA microarray and biosensor technologies, *Trends Biotechnol.* **23** (3): 143–149.

McGall G. H., Barone A. D., Diggelmann M., Fodor S. P. A., Gentalen E., and Ngo N., 1997, The efficiency of light-directed synthesis of DNA arrays on glass substrates, *J. Am. Chem. Soc.* **119** (22): 5081–5090.

Mougin K., Ham A. S., Lawrence M. B., Fernandez E. J., and Hillier A. C., 2005, Construction of a tethered poly(ethylene glycol) surface gradient for studies of cell adhesion kinetics, *Langmuir* **21** (11): 4809–4812.

Oh S. J., Ju J., Kim B. C., Ko E., Hong B. J., Park J.-G., Park J. W., and Choi K. Y., 2005, DNA microarrays on a dendron-modified surface improve significantly the detection of single nucleotide variations in the p53 gene, *Nucleic Acids Res.* **33** (10): e90.

Park S.H., and Krull, U.J., 2006, A spatially resolved nucleic acid biochip based on a gradient of density of immobilized probe oligonucleotide, *Anal. Chim. Acta* **564**: 133–140.

Pease A. C., Solas D., Sullivan E. J., Cronic M. T., Holmes C. P., and Fodor S. P. A., 1994, Light-generated oligonucleotide arrays for rapid DNA-sequence analysis, *Proc. Natl. Acad. Sci. USA* **91**: 5022–5026.

Peterson A. W., Heaton R. J., and Georgiadis R., 2000, Kinetic Control of Hybridization in Surface Immobilized DNA Monolayer Films, *J. Am. Chem. Soc.* **122**: 7837–7838.

Peterson A. W., Heaton R. J., and Georgiadis R. M., 2001, The effect of surface probe densityon DNA hybridization, *Nucleic Acids Res.* **29** (24): 5163–5168.

Pirrung M. C., Fallon L., and McGall G. H., 1998, Proofing of photolithographic DNA synthesis with 3', 5'-dimethoxybenzoinyloxycarbonyl-protected deoxynucleoside phosphoramidites, *J. Org. Chem.* **63** (2): 241–246.

Pirrung M. C., Wang L., and Montague-Smith M. P., 2001, 3'-nitrophenylpropyloxycarbonyl (NPPOC) protecting groups for high-fidelity automated 5' -> 3' photochemical DNA synthesis, *Org. Lett.* **3** (8): 1105–1108.

Piunno P. A. E., and Krull U. J., 2005, Trends in the development of nucleic acid biosensors for medical diagnostics, *Anal. Bioanal. Chem.* **381**: 1004–1011.

Piunno P. A. E., Watterson J. H., Kotoris C. C., and Krull U. J., 2005, Alteration of the selectivity of hybridization of immobilized oligonucleotide probes by co-immobilization with charged oligomers of ethylene glycol, *Anal. Chim. Acta* **534**: 53–61.

Plummer S. T., Wang Q., and Bohn P. W., 2003, Electrochemically Derived Gradients of the Extracellular Matrix Protein Fibronectin on Gold, *Langmuir* **19** (18): 7528–7536.

Shchepinov M. S., Case-Green S. C., and Southern E. M., 1997, Steric factors influencing hybridisation of nucleic acids to oligonucleotide arrays, *Nucleic Acids Res.* **25** (6): 1155–1161.

Steel A. B., Levicky R. L., Herne T. M., and Tarlov M. J., 2000, Immobilization of nucleic acids at solid surfaces: Effect of oligonucleotide length on layer assembly, *Biophys. J.* **79**: 975–981.

Terrill R. H., Balss K. M., Zhang Y., and Bohn P. W., 2000, Dynamic monolayer gradients: Active spatiotemporal control of alkanethiol coatings on thin gold films, *J. Am. Chem. Soc.* **122** (5): 988–989.

Watterson J. H., Piunno P. A. E., and Krull U. J., 2002, Towards the optimization of an optical DNA sensor: control of selectivity coefficients and relative surface affinities, *Anal. Chim. Acta* **457**: 29–38.

Watterson J. H., Piunno P. A. E., Wust C. C., and Krull U. J., 2000, Effects of Oligonucleotide Immobilization Density on Selectivity of Quantitative Transduction of Hybridization of Immobilized DNA, *Langmuir* **16** (11): 4984–4992.

Watterson J. H., Raha S., Kotoris C. C., Wust C. C., Charabaghi F., Jantzi S. C., Haynes N. K., Gendron N. H., Krull U. J., Mackenzie A. E., and Piunno, P.A.E., 2004, Rapid detection of single nucleotide polymorphisms associated with spinal muscular atrophy by use of a reusable fibre-optic biosensor, *Nucleic Acids Res.* **32** (2): e18.

Zeng J., Almadidy A., Watterson J. H., and Krull U. J., 2003, Interfacial hybridization kinetics of oligonucleotides immobilized onto fused silica surfaces, *Sens. Actuators B* **90**: 68–75.

BIOSENSING CONFIGURATIONS USING GUIDED WAVE RESONANT STRUCTURES

I. ABDULHALIM[*]
Department of Electrooptic Engineering
Ben Gurion University of the Negev
Beer Sheva 84104, Israel

Abstract. Resonant structures are characterized by a high quality factor representing the sensitivity to perturbations in a cavity. In guided wave resonant structures the optical field is evanescent, forming a region where the resonance can be modified by externally varying the refractive index within this evanescence region. The resonance nature of these structures then allows high sensitivity to analytes, gases, or other external index perturbations down to the order of 10-8 RIU. In this article several configurations of guided wave resonant structures and their use for sensing are reviewed with special emphasis on grating coupled resonant structures. The sensor performance is discussed using analytic approaches based on planar waveguide sensors theory and using the 4 × 4 characteristic matrix approaches for multilayered structure and with homogenized grating treated as a uniaxial thin film. The results agree very well with experiment and with rigorous electromagnetic calculations even when the cover is anisotropic medium such as a liquid crystal that can be used for tunable filtering or temperature sensing.

Keywords: Biosensors; optical sensors; tunable filters; resonant mirror; waveguides; gratings; guided wave resonance; microresonators

1. Introduction

Optical sensors have been under extensive research during the last two decades due to their wide variety of applications in biology, medicine,

[*] abdulhlm@bgu.ac.il

W.J. Bock et al. (eds.), Optical Waveguide Sensing and Imaging, 211–228.

environment and industry for monitoring of parameters such as temperature, concentration, pressure, biochemical and biological entities[1]. Sensors based on surface plasmon resonance (SPR) are being used for monitoring of fabrication processes of biochemical and pharmaceutical industries[2]. Fiber based sensors[3] are being used in concrete for stress monitoring, in machinery and industrial areas for monitoring of pressure and temperature over kilometers length scale. Optoelectronic sensors are used for imaging, profiling, auto-focusing and tracking. In the material side, metal nano-particles, nano-tubes, quantum dots and porous materials are under extensive studies for biosensing applications due to their unique optical properties. Guided wave structures such as planar waveguides and optical fibers allow optical sensing based on the interaction between the evanescent field outside the confinement region and the analyte to be sensed. Several important parameters define the quality of a sensor: (i) sensitivity, (ii) specificity, and (iv) reliability. In order to enhance the sensitivity, several techniques were developed such as the surface enhanced optical effects near a metal nano-particle[4], microresonators[5], resonant mirror waveguides[6], and other resonance effects in waveguides or other photonic structures[7].

One interesting configuration is the anti-resonant reflective optical waveguides (ARROW) which are waveguides with the light confinement acheived by Fabry-Perot anti-resonant reflectors, rather than total internal reflection (TIR). As a result of the light confinement mechanism, ARROWs can be constructed such that the light is confined in a low refractive index medium surrounded by high refractive index reflecting boundaries. ARROWs have been proposed as optical sensors[8]. This configuration uses the evanescent field outside the ARROW waveguide to perform sensing, leading to relatively low sensitivity.

Another interesting resonant structure is the newly introduced field of micro-resonators (MRs). In MRs, resonance in a transparent dielectric micro-resonator occurs when light, confined by TIR along the inside of the resonator surface, orbits near a recognition particle's surface and returns in phase after each revolution. The sensitivity is improved by several orders of magnitude due to the fact that the light interacts with the same analyte molecule captured by the recognition particle for several thousand times unlike single-pass techniques. The frequencies of the whispering gallery modes (WGMs), characterized by the number of wavelengths within an orbit, are extremely sensitive to added dielectric material on the recognition particle's surface. Just an atomic thickness can lead to a detectable shift of a specific resonance frequency[9]. Optical micro-resonators have attracted interest during the last few years in the biosensing community, due to (i) their small size requiring analyte solutions in nanoliter volumes, (ii) high quality factors, and (iii) unprecedented sensitivity. These tiny optical cavities, whose dimmers may vary from a few to several micrometers, deliver quality factors as large as 3×10^9 and beyond. Such enormously high quality factors represent unique performance characteristics: an extremely narrow resonant linewidth, long decay time, and a high energy density. Optical micro-resonators have various types of shapes such as cylindrical, spherical, spheroidal/toroidal, and ringlike. The underlying principle is the provision of efficient energy transfer to the resonant circular TIR guided wave, representing the WGM, through the evanescent field of a guided wave or a TIR spot in the coupler. Efficient coupling occurs upon fulfillment of two main conditions: (i) phase synchronization, and (ii) significant overlap of the WGM and the coupler mode. Different techniques have been demonstrated for coupling the light into the micro-resonators including the prism couplers, side-polished optical fiber-couplers, and "pigtailing" couplers utilizing angle-polished fiber tips in which a core-guided wave undergoes TIR.

Vollmer[10] et al. demonstrated the use of WGM microsphere biosensors for the detection of protein adsorption: the adsorption of a single layer of Bovine Serum Albumin (BSA) caused the wavelength to shift by approximately 16 ppm. An optical microsphere resonator used[11] for the detection of thrombin using aptamer as the recognition molecule delivered a detection limit on the order of 1 NIH Unit/ml (1 NIH unit of thrombin clots a standard fibrinogen solution in 15 s at 37°C). A multiplexed platform for DNA quantification was developed with two microsphere cavities evanescently coupled to the same single optical fiber, the sensitivity of this device being as high as 6 pg/mm^2 mass loading[12]. Boyd[13] et al. described the use of WGM disk micro-resonators for the detection of pathogens using selective recognition receptors, the devices being capable under optimum conditions of detecting as few as 100 molecules.

Another interesting resonant configuration is the resonant mirror (RM) waveguide sensor. The RM is effectively a prism coupler where the air gap has been replaced by a low-index dielectric layer. The RM device structure, usually consisting of a high-index substrate (n = 1.72), a thin low-index spacer (about 550 nm of silica) and a very thin monomode waveguiding layer (about 80 nm of Si_3N_4). The high-index resonant layer acts as both a waveguiding and a sensing layer. Light incident above the critical angle on the substrate-spacer interface is coupled into the waveguiding layer via the evanescent field in the spacer, when the propagation constants in the substrate and waveguide match. For monochromatic light, this occurs over a very narrow range of incidence angles, typically spanning considerably less than $1°$. Alternatively, it can be operated at a fixed incidence angle, and coupling occurs over a narrow range of wavelengths[14]. The RM sensor was developed for immunosensors (for example, the commercially available product IAsys from Affinity Sensors, a company in Cambridge, UK), as it is very sensitive to changes in the refractive index of the interfacial layer caused by the binding of macromolecules such as proteins to immobilized biorecognition species such as antibodies[15].

Resonant grating waveguide structures (GWS) or guided mode resonance (GMR) structures have also been used for biosensing. They are very sensitive to the adsorption/desorption of molecules on the waveguide surface and to any change of refractive index of the medium covering the surface of the sensor chip. When the GMR structure is illuminated with an incident light beam, the diffracted light matches the guided-mode condition and interference with the zero order beams causes resonant reflection backwards. This happens at a specific wavelength and incidence angle of the incident beam at which the resonance condition is satisfied, whereby the re-diffracted beam destructively interferes with the transmitted beam, so that the incident light beam is completely reflected[16,17,18]. In this article I shall review the resonant waveguide configurations used to enhance the sensitivity of biosensors in general and in particular will give more details of the guided wave resonance structures.

2. Evanescent Wave Sensing

Most of the optical sensing techniques are based on the existence of evanescent wave in the region where the analyte to be sensed is located. Examples are: TIR, ATR, SPR, fibers and waveguides, LSPR, micro-resonators, grating waveguide resonant structures, and resonant mirror sensors. Evanescent waves arise when there is a confinement region in which the majority of the optical density exists, however outside this region a tail of the optical field exists forming the evanescent wave. Figure 1 shows a general schematic of the confinement region and the two bounding regions called substrate and cover or analyte.

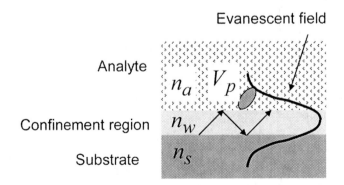

Figure 1. Schematic of evanescent wave sensor based on planar coptical waveguide.

The dielectric function of the structure maybe written as:

$$\varepsilon = \begin{cases} n_w^2 & r \in V_w \\ n_{a,s}^2 & r \notin V_w \end{cases},$$
(1)

where n_w, n_a, n_s are the refractive indices in the confinement (waveguide) region, the analyte (cover) region and the substrate region with V_w being the space volume of the confinement region. Assuming a particle with refractive index n_p is added to the analyte and caused a variation $\delta\varepsilon = n_a^2 - n_p^2$ in the dielectric function of the analyte within the volume of the particle V_p. The wave equation before the addition of the particle is:

$$\nabla x \nabla x E_i = k_i^2 \varepsilon E_i,$$
(2)

where E_i, k_i are the electric field and the wave vector before the addition of the particle. After the addition of the particle, the electric field, the dielectric function and the wave vector changes to: E_f, $\varepsilon_f = \varepsilon + \delta\varepsilon$, $k_f = k_i + \delta k$ so that the wave equation becomes:

$$\nabla x \nabla x E_f = k_f^2 \varepsilon_f E_f,$$
(3)

Multiplying by E_i^*, subtracting equation (2) from equation (3) and integrating over the entire volume leads to:

$$(k_i^2 - k_f^2) \int E_f \varepsilon E_i^* dr = k_f^2 \int_{V_p} E_i^* \delta\varepsilon E_f dr$$
(4)

Using first order perturbation theory in δk we get:

$$\delta k \approx \frac{k_i}{2} \frac{\int_{V_p} \delta\varepsilon E_i^* E_p dr}{\int_V \varepsilon E_i^* E_i dr}$$
(5)

Hence the dielectric perturbation in the evanescence region caused a shift in the guided wave vector determined by the overlap integral normalized to the mode energy integral. This is the essence of evanescent wave sensing. To demonstrate this, assume a single spherical dielectric particle is added, then the electric field within the volume that the particle occupying is given by:

$$E_p = \frac{3n_a^2}{2n_a^2 + n_p^2} E_a$$
(6)

Inserting this into equation (5) leads to:

$$\frac{\delta k}{k_i} \approx -\frac{3}{2}\frac{1-(n_p/n_a)^2}{2+(n_p/n_a)^2}\frac{V_p}{V_{int}} \tag{7}$$

Where here V_{int} is the interaction volume between the particle and the evanescent field defined as the region where the field E_i extends. It should be noted that the ratio $\delta k / k_i$ can be identified with the relative change in the mode effective index $\delta n_{eff} / n_{eff}$. Figure 2 shows the behavior of the relative change in the effective index with the ratio n_p / n_a in units of V_p/V_{int}. Note that as V_{int} is smaller, that is the field is localized, the mode index variation becomes more sensitive to the particle index, a fact that might have some relation to the surface enhancement of optical effects near metallic nano-particles. When the particle has a complex dielectric constant $\varepsilon_p = n_p^2$ then an enhanced shift is obtained when $Re\{2\varepsilon_a + \varepsilon_p\} = 0$, which again explains the enhanced sensing of metallic nano-particles in the evanescent region.

The sensitivity of the evanescent wave sensors, in particular the guided wave ones, is defined by:

$$S = \frac{\partial n_{eff}}{\partial n_a} \tag{8}$$

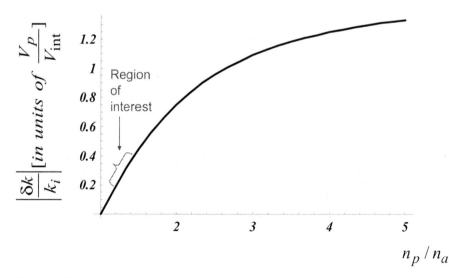

Figure 2. Variation of the relative change in the mode effective index versus the ratio between the particle index to the analyte index.

The mode effective index is a function of many parameters:

$$n_{eff} = f(n_a, n_w, n_s, n_0, n_1, d_g, d_w, \gamma_i, \lambda, \Lambda, l, \rho), \tag{9}$$

where the additional parameters here are $n_0, n_1, d_g, d_w, \gamma_i, \Lambda, \lambda, l, \rho$, defined as: the refractive indices of the two regions forming the rectangular grating in case of grating coupling, the grating height, the waveguide thickness, the incidence angle, the grating period, the wavelength, the mode number and mode type with $\rho = 0$ for TE and $\rho = 1$ for TM. The equations describing the sensitivity of planar waveguides were derived by Tiefenthaler and Lukosz[19] for the three layer waveguide and can be written as follows:

$$S = \frac{n_a \delta_a}{n_{eff} d_{eff}} \left[\frac{n_w^2 - n_{eff}^2}{n_w^2 - n_a^2} \right] \left[2 \frac{n_{eff}^2}{n_a^2} - 1 \right]^\rho \tag{10}$$

Where $d_{eff} = d_w + \delta_a + \delta_s$ an effective thickness with $\delta_{s,a}$ being the penetration depths in the substrate and analyte given by:

$$\delta_{s,a} = \frac{1-\rho}{k\sqrt{n_{eff}^2 - n_{s,a}^2}} + \frac{\rho \left[(n_{eff}/n_w)^2 + (n_{eff}/n_{s,a})^2 - 1 \right]^{-1}}{k\sqrt{n_{eff}^2 - n_{s,a}^2}} \tag{11}$$

This defines the extent of the evanescent field inside the substrate or the analyte:

$$E_l(z) = E_l(z_{s,a}) \exp(-z / \delta_{s,a}), \tag{12}$$

with $z_{s,a}$ defining the waveguide-substrate or the waveguide-analyte interfaces. To determine the sensitivity to the analyte index from equation (10) the mode effective index has to be determined which is a function of n_a itself. Therefore the sensitivity determination requires first a solution of the mode index dispersion relation determined by the phase matching condition:

$$l\pi = kd_w\sqrt{n_w^2 - n_{eff}^2} - arctan\left[\left(\frac{n_w}{n_s} \right)^{2\rho} \left(\frac{n_{eff}^2 - n_s^2}{n_w^2 - n_{eff}^2} \right)^{0.5} \right] -$$

$$arctan\left[\left(\frac{n_w}{n_s} \right)^{2\rho} \left(\frac{n_{eff}^2 - n_a^2}{n_w^2 - n_{eff}^2} \right)^{0.5} \right] \tag{13}$$

Few consequences can be understood based on these analytic expressions:

- When $n_a \approx n_w$ the analyte is part of the WG and the sensitivity is high.
- When $n_a \approx n_{eff}$ the penetration depth in the analyte region is high and so the sensitivity is high.
- When $n_a < n_s$ there exists a WG thickness just above the cutoff where the sensitivity is a maximum. At the cutoff $S = 0$, $n_{eff} = n_s$ because the energy is in the substrate.
- When $n_a > n_s$ the maximum sensitivity is at the cutoff because the field is in the analyte and nothing in the substrate.
- At large d_w, $S \rightarrow 0$ because the energy becomes confined in the waveguide.

3. Grating Coupled Resonant Structure

Sharp resonances in the diffraction efficiency of diffraction gratings can be traced back to 1902, the so called Wood anomaly[20]. Distinction between the resonant and nonresonant anomalies was first proposed in 1941 by Fano[21] who found that the former is because of the excitation of guided waves and the latter appearing when some diffraction order is being passed off. In 1965, Hessel and Oliner[22] proposed a phenomenological approach to resonant anomalies that introduces the poles and the zeros of the diffraction efficiency. The pole appears because of guided-wave excitation which is a result of the solution of the homogeneous problem when a guided wave exists without an incident wave. This solution requires that the scattering matrix that links the diffracted- and the incident-field amplitudes has a zero determinant. In so far as the diffracted amplitudes are inversely proportional to this determinant, they have a singularity, i.e., a complex pole, which equals to the guided wave propagation constant. Because of energy-balance and continuity requirements, this pole must be accompanied by a zero of the amplitudes of the propagating diffraction orders. The values of the poles and the zeros are complex, and their positions in the complex plane depend on grating parameters but not on the angle of incidence. The phenomenological approach (as well as grating anomalies, in general) has been the subject of extensive studies. Several reviews[23,24] can be found that describe this approach and show how to use its results for predicting the behavior of anomalies. Recently the subject was again revived[16,17,25,26] in connection with dielectric-grating anomalies when such gratings are used as narrow-band optical filters. In brief, when a waveguide mode is excited in a dielectric grating (usually a corrugated waveguide) the pole leads to a peak and the zero to a dip in the diffraction efficiency and, in particular, in the reflectivity and the transmittivity of the device. When the overall (non-resonant) reflectivity is low the high (theoretically 100%) and narrow peak in the reflectivity can be used for spectral filtering[27,28]. Since the propagation constants of the guided wave are polarization dependent, the position of the peak depends strongly on the polarization; thus the filtering properties are polarization selective.

Guided-mode resonance (GMR) is a peculiar diffraction phenomenon of waveguide gratings with definite parameters and incident light conditions. It refers to a sharp peak in the diffraction efficiency spectrum of waveguide gratings. At resonance, efficient energy exchange between the reflected and transmitted waves occurs in small parameter ranges (for example, wavelength, angle of incidence, or refractive index). Physically, this is due to coupling of the externally propagating diffracted fields to the modes of the waveguide. For a sub-wavelength grating, the grating period is shorter than the incident wavelength, only the zero-order forward and backward diffracted waves propagate, while all higher order waves are cut off. High reflection mirrors, filters and polarization devices, which are widely used in the fields of lasers, optical communication and optoelectronics, can be realized by using the properties of high diffraction efficiency and narrow linewidth of GMR. Moreover, the applications of GMR in biology[29], sensors[30,31], and medicine[32] have also attracted people's attention. There are many reports of theory and experiments on GMR, which prove the correctness of GMR as well as the feasibility of manufacture. Experimental results verifying the theoretically predicted high resonant efficiencies for reflection filters have also been reported in the millimeter wave region[33] in the microwave region[34], in the near infrared region[35] and in the visible regions[17].

3.1. THE CONDITION FOR THE GMR

The basic structure of the GMR device is shown in Figure 3 where the grating layer is on top of the waveguide layer and the top layer could be the analyte material filling both the spaces between the grating lines and the space above the gratings. A cavity is formed[36,37] for the diffracted order and a reflection resonance is obtained when the phase difference between the transmitted and reflected waves is a multiple of π. To show this we start from the grating equation:

$$n_0 k \sin \gamma_i + mG = n_w k \sin \gamma_d, \qquad (14)$$

where $G = 2\pi / \Lambda$, is the grating vector. When the diffracted beam of order m becomes a guided mode with an effective mode index

$$n_{eff} = n_0 \sin\gamma_i + m\lambda/\Lambda \qquad (15)$$

The phase difference between the transmitted and reflected waves is:

$$\varphi_{t-r} = \varphi_{0-w} + \varphi_{TIR} + 2\varphi_{diff}, \qquad (16)$$

where the phase difference due to pathlength difference is: $\varphi_{0-w} = 2k_w d_w$ with $k_w = kn_w$ and the phase difference $\varphi_{diff} = \varphi_{Fresnel} - \pi/2$ is due to diffraction and Fresnel reflection at the interfaces. Substituting all this into equation (16) yields:

$$\varphi_{t-r} = 2k_w d_w + \varphi_{TIR} + 2\varphi_{Fresnel} - \pi \qquad (17)$$

The guided wave condition is:

$$2k_w d_w + \varphi_{TIR} + 2\varphi_{Fresnel} = 2\pi l \qquad (18)$$

Combination of equations (17) and (18) leads to:

$$\varphi_{t-r} = \pi(2l - 1) \qquad (19)$$

When the diffracted beam is a guided wave, destructive interference occurs between the transmitted and reflected beams leading to resonance in reflection.

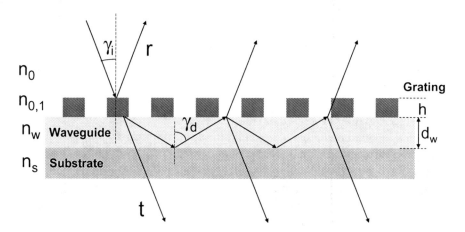

Figure 3. Schematic of the guided wave resonant structure.

3.2. THE REFLECTION PEAK SHAPE

The reflected resonant peak shape was shown by Nevier et al.[38] to be a Lorentzian. The angular shape of the peak can be written as:

$$R = \frac{|\kappa_a|^2 / k}{\left(n_0 \sin\gamma_i - n_0 \sin\gamma_{ipeak}\right)^2 + (\Gamma / k)^2}, \tag{20}$$

where here κ_a represents a coupling constant and Γ is a loss parameter. Note that γ_i here is the incidence angle in the medium above the grating of index n_0, while if the light is incident from air on this medium then in terms of the incidence angle in air γ_a, the expression $n_0 \sin\gamma_i = \sin\gamma_a$ should be replaced with $\sin\gamma_a$. The peak location is determined by equation (15): $n_0 \sin\gamma_{ipeak} = n_{eff} - m\lambda / \Lambda$ while the width at half the maximum is: $\Delta\gamma_i = (180 / \pi)\left(\lambda\Gamma / (\pi \cos(\gamma_{ipeak}))\right)$. The spectral shape maybe written as:

$$R = \frac{|\kappa_w|^2 (\Lambda\lambda_{peak} / 2\pi)^2}{\left(\lambda - \lambda_{peak}\right)^2 + \lambda^2 (\Lambda\Gamma / 2\pi)^2}, \tag{21}$$

where the peak wavelength is determined by equation (15): $\lambda_{peak} = (n_{eff} - n_0 \sin\gamma_i)\Lambda / m$ while the spectral width is given by: $\Delta\lambda = (\lambda_{peak}\Lambda\Gamma / \pi)$. Note that R=1 when $\kappa_a = \kappa_w = \Gamma$.

3.3. DESIGN CONSIDERATIONS

The basic parameters for the design of GMR structure can be determined from the equations in the previous section particularly the peak position, shape and width. The effective index however should be determined from the mode dispersion relation similar to the three layer WG problem described by equation (13). Since the grating layer is much thinner than a wavelength it can be ignored and the results in this approach are obtained in good approximation. Alternatively one can use more rigorous electromagnetic calculation such as the use of the eigen-functions approach, the rigorous coupled wave approximation (RCWA), the Fourier approach or the scattering matrix approach[39]. These approaches can give the resonance spectrum including absorption, exact value of the peak width and its dependence on the grating parameters. A less heavy approach uses the characteristic matrix approach where the grating layer is homogenized to a uniaxial thin film within the effective medium approximations. The 4 × 4 matrix approach can handle anisotropic layers and it was used recently by the present author[18] to show that the effective mode index calculated this way agrees very well with the rigorous approaches. In order to maximize the peak reflectivity, the grating period should be chosen less than the wavelength so that only the zero order is supported and the first order diffraction exists in the WG (m=1). The existence of higher modes will decrease the diffraction efficiency and pull part of the energy away into the higher orders. Losses are a result of absorption, scattering due to imperfections particularly in the WG layer where the interaction region is large and due to imperfect collimation of the incident light beam. As a sensor, the WG index and thickness should be chosen so that the evanescent field extends more in the analyte region. In order to reduce the background reflection outside the resonance region, care should be taken to the design of the layers and perhaps the inclusion of anti-reflection coating (ARC) in between. As this is not so easy with the rigorous approaches due to the heavy numerical calculation, optimization can be done with thin film design software's or the use of the characteristic matrix approach with the grating film homogenized to uniform uniaxial film. Fine tuning of the structure parameters can then be done with the rigorous calculation.

3.4. THE GMR AS A SENSOR

There are several attractive properties of the GMR to be used both as a narrow filter and as a sensor: (i) planar geometry (ii) made of standard dielectric materials (iii) can be manufactured easily in mass production with Si fabrication technology on the wafer scale and used for multi-sensing functionality (iv) can be operated at normal incidence (v) exhibits large sensitivity, at least comparable to the sensitivity pf the planar WG sensor and (vi) can be operated both in spectral mode and in angular mode. Figure 4a shows the angular and spectral operation modes of the GMR device. In the angular mode, a single wavelength is used and a beam with a spread of angles, for example the natural spread from a laser diode. The center of mass of the beam is detected using an array of detectors such as a CCD camera. Any shift in the reflection resonant angle will affect the center of mass of the beam. In the spectral mode a collimated beam is used containing a relatively wide spectral range and the spectrum is analyzed using a spectro-meter. Alternatively a tunable source can be used for continuous scanning of the wavelength and a single pixel detector. In figure 4b the normal incidence operation mode is illustrated which is usually preferable in particular when multi-sensing using an array of GMR structures is required.

As can be seen from the gratings equation, the spectral sensitivity is:

$$(\partial \lambda_{peak} / \partial n_a) = (\Lambda / m)(\partial n_{eff} / \partial n_a),$$

therefore we can conclude that the sensitivity is determined by the sensi-ivity of n_{eff} in a similar fashion to the sensitivity of a planar waveguide. The largest sensitivity is obtained for the first order diffraction m=1 and for larger Λ. Note that the sensitivity in the angular mode is slightly less because $sin \gamma_i < \gamma_i$ except for small angles, where it becomes comparable.

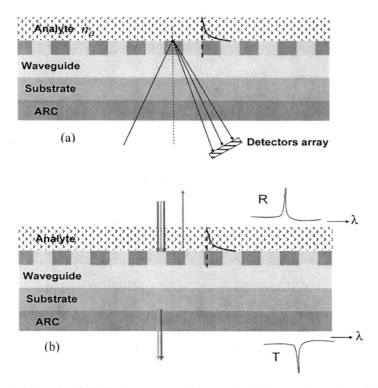

Figure 4. Schematic of GMR structure as a biosensor in (a) the angular mode and (b) the spectral mode at normal incidence.

The combination of a colorimetric resonant grating and photonic crystal embedded in the plastic surfaces of microtiter plates (96-, 384-, and 1536-well) has been developed by SRU Biosystems as a label-free, high-throughput, screening platform. The sensor can detect a shift in wavelength as low as half a picometer. Binding interactions can be quantified with proteins, cells, and small molecules. Sensitivity is quoted in the 0.05 µg/ml to 1 mg/ml range with molecular weights < 200 Da. Corning has also developed a label-free detection platform that contains resonant GWS in the bottoms of 384-well microtiter plates. When illuminated with broadband light, the optical sensors inside each well reflect only a specific wavelength that is a sensitive function of the index of refraction close to the sensor surface. The platform has a sensitivity of 5 pg/mm^2, which enables the detection of the binding of a 300-Da molecule to a 70-kDa immobilized molecule[29-31].

As an example of a design for water sensor operating at normal incidence we considered a grating of pitch $\Lambda = 500\ nm$, having lines of height $h = 100\ nm$ and index $n_1 = 3.6$ while the spaces are filled by the liquid analyte of index around $n_a = 1.33$ corresponding to water. The waveguide layer has an index and thickness of $n_w = 1.6, d_w = 500\ nm$. The sensitivity $(\partial n_{eff} / \partial n_a)$ calculated from the slope is 0.21 for the TM0 mode and 0.24 for the TE0 mode which is comparable to the maximum sensitivity reported for planar waveguides when $n_a < n_s$. The spectral sensitivity is then equals: $\Lambda(\partial n_{eff} / \partial n_a) \approx 100 - 120\ nm / RIU$, hence if the system minimum spectral detectability is 1pm, one can measure index variations of the order of $10^{-5}\ RIU$. For analytes with $n_a > n_s$ and thin waveguide layer the sensitivity can be enhanced by few times as expected.

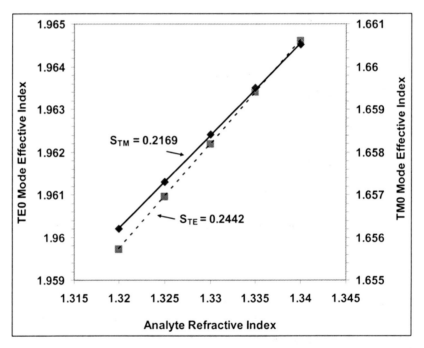

Figure 5. Calculated sensitivity of water sensor for the zero order modes. Parameters of the GMR structure are given in the text.

One of the potential applications of the GMR structure is for tunable filtering and temperature sensing using a liquid crystal (LC) that exhibits large electro-optic and thermo-optic effects.

In Figure 6a, a simplified design is shown where the LC itself is the waveguide on top of the grating. The LC has the refractive indices and tilt angle: $n_\perp = 1.611, n_{||} = 1.830, \theta = 90°$. The grating refractive indices, height and fill factor are:

$$n_H = 1.95, n_L = 1.6, d_g = 0.1\Lambda, f = 0.5.$$

Figure 6b shows comparison of calculated mode effective index using the analytic 4 × 4 matrix approach[18,40] and using the rigorous coupled wave approach (RCWA).[39]

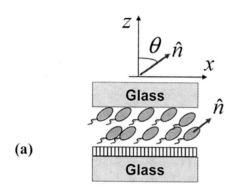

(a)

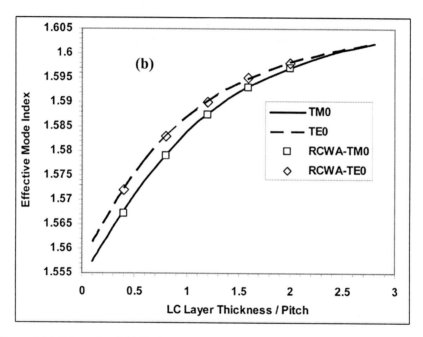

Figure 6. (a) Schematic of GMR structure with the LC layer acting as the waveguide layer on top of the grating sandwiched between two glass plates. (b) The effective mode index versus the normalized LC layer thickness calculated both using the rigorous RCWA approach and using the analytic 4 × 4 matrix approach. Structure parameters are given in the text.

The agreement is excellent, hence confirming that it is possible to use the 4 × 4 matrix approach for the design of GMR structures. One of the advantages of using liquid crystals is the possibility of tuning the resonance using an ac voltage of few volts applied between the two glass plates.

The voltage causes rotation of the molecules towards the normal to the plates hence decreasing the tilt angle θ which in turn causes variation in the effective index.

4. Conclusions

Among the sensors based on the resonant guided wave structures, the grating coupled resonant structure is one of the most attractive ones as it allows normal incidence operation in spectral or angular modes with high sensitivity. The resonance in the GMR can be observed as a peak in the reflectivity either observed at fixed incidence angle versus wavelength or at fixed wavelength versus angle. It is shown that the design follows the same rules as for planar waveguide structures and the use of the characteristic matrix is proposed as a powerful tool when multilayered structures are involved. With the grating homogenized as a uniaxial plate the use of the 4×4 matrix approach is necessary instead of the 2×2 matrix method where the mode effective indices are determined as the poles of the reflectivity function. The GMR structure has the potential of being used in muti-arrays as a biochip using standard Si technology.

References

1. F. S. Ligler and C. A. Rowe Taitt, Eds., *Optical Biosensors: Present and Future*, Elsevier, Amsterdam, The Netherlands, 2002.
2. J. Homola, S. S. Yee, and G. Gauglitz, Surface plasmon resonance sensors: review, *Sensors and Actuators B* **54**, 3–15 (1999).
3. L. S. Grattan and B. T. Meggett, Eds., *Optical Fiber Sensor Technology: Advanced Applications – Bragg Gratings and Distributed Sensors*, Kluwer Acdemic Publishers, Boston, 2000.
4. D. A. Genov, A. K. Sarychev, V. M. Shalaev, and A. Wei, Resonant field enhancements from metal nanoparticle arrays, *Nano Letters* **4**, 153–158 (2004).
5. A. B. Matsko, A. A. Savchenkov, D. Strekalov, V. S. Ilchenko, and L. Maleki, Review of applications of whispering-gallery mode resonators in photonics and nonlinear optics, *IPN Progress Report 42–162* August 15, (2005)
6. R. Cush, J. M. Cronin, W. J. Stewart, C. H. Maule, J. Molloy, N. J. Goddard, The resonant mirror: a novel optical biosensor for direct sensing of biomolecular interactions Part I: Principle of operation and associated instrumentation, *Biosensors and Bioelectronics* **8**, 347–354 (1999).
7. I. D. Block, L. L. Chan, and B. T. Cunningham, Photonic crystal optical biosensor incorporating structured low-index porous dielectric, *Sensors and Actuators B* **120**,187–193 (2006).

8. M. A. Duguay, Y. Kokobun and T. L. Koch, Antiresonant Reflecting Optical Waveguides in SiO2-Si multilayer structures, *Applied Physics Letters* **49**, 13–16 (1986), 13–16.

9. K. Vahala, Ed., *Optical Microcavities*, World Scientific, Singapore, 2004.

10. F. Vollmer, D. Braun, A. Libchaber, M. Khoshsima, I. Teraoka, and S. Arnold, Protein detection by optical shift of a resonant microcavity, *Applied Physics Letters* **80**, 4057–4059 (2002).

11. H. Zhu, J. D. Suter, I. M. White, and X. Fan, Aptamer based microsphere biosensor for thrombin detection, *Sensors* **6**, 785–795 (2006).

12. F. Vollmer, S. Arnold, D. Braun, I. Teraoka, and A. Libchaber, Multiplexed DNA quantification by spectroscopic shift of two microsphere cavities, *Biophysical Journal* **85**, 1974–1979 (2003).

13. R. W. Boyd and J. E. Heebner, Sensitive disk resonator photonic biosensor, *Applied Optics* **40**, 5742–5747 (2001).

14. N. J. Goddard, D. Pollard-Knight, and C. H. Maule, Real-time biomolecular interaction analysis using the resonant mirror sensor, *The Analyst* **119**, 583–588 (1994).

15. A. Brecht, A. Klotz, C. Barzen, G. Gauglitz, R. D. Harris, G. R. Quigley, J. S. Wilkinson, P. Sztajnbok, R. Abukensha, J. Gascon, A. Oubina, and D. Barcelo, Optical immunoprobe development for multiresidue monitoring in water, *Anal. Chim. Acta* **362**, 69–79 (1998).

16. A. Sharon, D. Rosenblatt, A. A. Friesem, H. G. Weber, H. Engel, and R. Steingrueber, Light modulation with resonant grating-waveguide structures, *Optics Letters* **21**, 1564–1566 (1996).

17. A. Sharon, D. Rosenblatt, and A. A. Friesem, Resonant grating–waveguide structures for visible and near-infrared radiation, *J. Opt. Soc. Am. A* **14**, 2985–2993 (1997).

18. I. Abdulhalim, Anisotropic layers in waveguides for tuning and tunable filtering, *Proceedings of SPIE* **6135**, 179–188 (2006).

19. K. Tiefenthaler, and W. Lukosz, Sensitivity of grating couplers as integrated optical chemical sensors, *J. Opt. Soc. Am. B* **6**, 209–220 (1989).

20. R. Wood, On a remarkable case of uneven distribution of light in a diffraction grating spectrum, *Philos. Mag.* **4**, 396–402 (1902).

21. U. Fano, The theory of anomalous diffraction gratings and of quasi-stationary waves on metallic surfaces (Sommerfeld's waves), *J. Opt. Soc. Am. A* **31**, 213–222 (1941).

22. A. Hessel and A. A. Oliner, A new theory of Wood's anomalies on optical gratings, *Appl. Opt.* **4**, 1275–1297 (1965).

23. M. Neviere, "The homogeneous problem," in *Electromagnetic Theory of Gratings*, R. Petit, ed. (Springer-Verlag, Berlin, 1980), Chap. 5.

24. E. Popov, "Light diffraction by relief gratings: a microscopic and macroscopic view," in *Progress in Optics*, E. Wolf, ed. (Elsevier, Amsterdam, 1993), Vol. XXXI, 139–187.

25. T. Tamir and S. Zhang, Resonant scattering by multilayered dielectric gratings, *J. Opt. Soc. Am. A* **14**, 1607–1616 (1997).

26. S. M. Norton, G. M. Morris, and T. Erdogan, Experimental investigation of resonant-grating filter line shapes in comparison with theoretical models, *J. Opt. Soc. Am. A* **15**, 464–472 (1998).

27. R. Magnusson and S. S. Want, New principles of optical filters, *Appl. Phys. Lett.* **61**, 1022–1024 (1992).

28. S. Peng and G. Morris, Experimental demonstration of resonant anomalies in diffraction from two-dimensional gratings, *Opt. Lett.* **21**, 549–551 (1996).

29. D. Wawro, S. Tibuleac, R. Magnusson, and H. Liu, Optical fiber endface biosensor based on resonances in dielectric waveguide gratings, *Proc. SPIE* **39**, 86 (2000).

30. B. Cunningham, P. Li, B. Lin, and J. Pepper, Colorimetric resonant reflection as a direct biochemical assay technique, *Sensors Actuators B* **81**, 316–328 (2002).
31. J. J. Wang, L. Chen, S. Kwan, F. Liu, and X. Deng, Resonant grating filters as refractive index sensors for chemical and biological detections, *J. Vacuum Science & Technology B: Microelectronics and Nanometer Structures*, 23, 3006–3010 (2005).
32. M. A. Cooper, Optical biosensors in drug discovery, *Nat. Rev. Drug Discovery*, **1**, 515–528 (2002).
33. V. V. Meriakri, I. P. Nikitin, and M. P. Parkhomenko, Frequency-selective properties of modified dielectric gratings, Int.J.Infrared & Millimeter Waves, **17**, 1769–1778 (1996).
34. R. Magnusson, S. S. Wang, T. D. Black and A. Sohn, Resonance properties of dielectric waveguide gratings:theory and experiments at 418 GHz, *IEEE Trans. Antennas Propag.*, **42**, 567–569 (1994).
35. P. S. Priambodo, T. A. Maldonado and R. Magnusson, Fabrication and characterization of high-quality waveguide-mode resonant optical filters, *Appl. Phys. Lett.* **83**, 3248–3250 (2003).
36. D. Rosenblatt, A. Sharon, and A. A. Friesem, Resonant Grating Waveguide Structures, *IEEE J. Quant. Electron.* **33**, 2038–2059 (1997).
37. S. Glasberg, A. Sharon, D. Rosenblat, and A. A. Friesem, Spectral shifts and line shapes asymmetries in the resonant response of grating waveguide structures, *Opt.Commu.* **145**, 291–299 (1998).
38. M. Nevie`re, R. Petit, and M. Cadilhac, Systematic study of resonances of holographic thin-film couplers, *Opt. Commun.* **9**, 48–53 (1973).
39. M. Nevière and E. Popov, *Light Propagation in Periodic Media: Differential Theory and Design*, Marcel Dekker, New-York, 2003.
40. I. Abdulhalim, Analytic propagation matrix method for linear optics of arbitrary biaxial layered media, *J. Opt. A: Pure Appl. Opt.*, **1**, 646–653 (1999).

POLARIZED LIGHT TRANSPORT INTO SCATTERING MEDIA USING A QUATERNION-BASED MONTE CARLO

JESSICA C. RAMELLA-ROMAN
The Catholic University of America, 620 Michigan Avenue NE, Washington, DC 20064

Abstract. Polarized light transport into a scattering media can be modeled using polarization sensitive Monte Carlo programs. This chapter will illustrate one such programs based on quaternion algebra. In the program the polarization reference plane is tracked using two unit-vectors *u* and *v*, quaternions are used to accomplish the rotation of the polarization reference plane. Comparison with Adding Doubling models showed that our Monte Carlo algorithm yields results with less than 1% error.

Keywords: Polarization, Quaternion, Monte Carlo

1. Introduction

Polarized light Monte Carlo programs are becoming of common use in biomedical optics [1, 2, 3]. Recently polarized light transport into biological media has been the center of several studies, including the enhancement of borders of skin cancer [4, 5], the discrimination of precancerous versus normal cells [6], and the study of tissue birefringence through polarized OCT [7]. Hence, Monte Carlo models that consider the polarization information of the scattered light are important to enhance our understanding of polarized light travel into biological media. Many of these Monte Carlo modeling tools are derived from the atmospheric optics and oceanography community. In fact, the first polarization sensitive Monte Carlo program was proposed by Kattawar and Plass [8, 9] to study sun light transmission through clouds and haze. Bruscaglione et al. measured the depolarization of a light pulses transmitted through a turbid media [10] and Bianchi et al. [11,12] developed a Monte Carlo program for nebulae analysis, finally Martinez and Maynard [13,14] studied the Faraday effect

W.J. Bock et al. (eds.), Optical Waveguide Sensing and Imaging, 229–241.

into an optically active medium. In the biomedical optics community polarization sensitive Monte Carlo methods have been proposed by several groups. The first paper on the topic was by Bartel and Hielsher [2] in 1999 although Schmitt et al. [15] in 1997 had shown models of linearly and circularly polarized light into scattering media. Rakovic et al. showed an extensive experimental and numerical validation of a polarized Monte Carlo program [16] and Cote' et al. [1] developed a Monte Carlo to study optically active molecules in turbid media. Wang et al. [17] used polarized Monte Carlo algorithms to simulate light propagation in birefringent media. Ramella-Roman et al. [3,18] showed three different ways to implement polarized Mont Carlo programs, one based on meridian planes tracking, one based on Euler angles rotations, and one based on quaternions rotations. Finally Xu et al. [19] proposed an algorithm that differs from all previous efforts. Instead of observing the propagation of the Stokes vector in the scattering media, they tracked the light electric field. Hence, their program is suitable for speckle and coherence studies. In this chapter we will illustrate the typical structure of a polarized Monte Carlo program and show some typical applications.

2. Implementation of a Polarization Sensitive Monte Carlo Program

2.1. MONTE CARLO

Monte Carlo is a computational algorithm that uses a stochastic approach to model a physical event [20]. A typical Monte Carlo program describing light travel into a scattering media is composed of four basic steps [21] *launch*, *hop*, *drop*, and *spin*. These steps are used to track the photon position, the deposition of energy at specific locations, as well as the scattering of the photon from one location to the next. When the polarization of the light traveling into the scattering media is taken into account, more steps must be added to the program. First the polarization reference system must be tracked; secondarily the choice of azimuth and scattering angles is accomplished using a rejection method, in lieu of obtaining the scattering angles directly from the phase function.

2.2. STOKES VECTORS AND SCATTERING MATRICES

In most Monte Carlo models the polarization of a beam traveling into an absorbing and scattering media, is characterized by a Stokes vector. The four dimensional Stokes vector completely describe the polarization of an electric field E [22] and is defined as:

$$S_0 = \langle E_x E_x^* \rangle + \langle E_y E_y^* \rangle$$
$$S_1 = \langle E_x E_x^* \rangle - \langle E_y E_y^* \rangle$$
$$S_2 = \langle E_x E_y^* \rangle + \langle E_y E_x^* \rangle \qquad (1)$$
$$S_3 = i\left(\langle E_x E_y^* \rangle - \langle E_y E_x^* \rangle\right)$$

The angular brackets are the time average. Examples of Stokes vectors are shown in Eq. 2.

$$\begin{bmatrix} 1 & 1 & 0 & 0 \end{bmatrix}^T \quad \textit{Linearly Polarized (Horizontal)}$$

$$\begin{bmatrix} 1 & -1 & 0 & 0 \end{bmatrix}^T \quad \textit{Linearly Polarized (Vertical)}$$

$$\begin{bmatrix} 1 & 0 & 1 & 0 \end{bmatrix}^T \quad \textit{Linearly Polarized (+45)} \qquad (2)$$

$$\begin{bmatrix} 1 & 0 & 0 & 1 \end{bmatrix}^T \quad \textit{Right Circularly Polarized}$$

From the equations above it is evident that the polarization of a field has no significance without the definition of a polarization reference plane; the tracking of this reference plane is a major concern in the Monte Carlo construction.

An optical system can be described in terms of Stokes and Mueller matrices [22], where the latter completely characterizes the polarizing properties of an optical element. Similarly the scattering of a photon from a well characterized particle can be accomplished with the use of its scattering matrix $M(\alpha)$, where α is the scattering angle. In the case of spherical scatterers many of the terms in the scattering matrix are zero due to symmetry, Eq. 3.

$$M(\alpha) = \begin{bmatrix} s_{11}(\alpha) & s_{12}(\alpha) & 0 & 0 \\ s_{12}(\alpha) & s_{11}(\alpha) & 0 & 0 \\ 0 & 0 & s_{33}(\alpha) & s_{34}(\alpha) \\ 0 & 0 & -s_{34}(\alpha) & s_{33}(\alpha) \end{bmatrix} \qquad (3)$$

The terms s_{11}, s_{12}, s_{33}, and s_{34} are the matrix scattering coefficients [23]

$$S_{11} = \frac{1}{2}\left(\left|S_2\right|^2 + \left|S_1\right|^2\right)$$

$$S_{12} = \frac{1}{2}\left(\left|S_2\right|^2 - \left|S_1\right|^2\right)$$

$$S_{33} = \frac{1}{2}\left(S_2^*S_1 + S_2 S_1^*\right) \tag{4}$$

$$S_{34} = -\frac{i}{2}\left(S_2^*S_1 - S_2 S_1^*\right)$$

S_1 and S_2 are α dependent terms (emitted in Eq. 4) and are functions of the size parameter x and the index of refraction n of the scattering particle and the surrounding media. In our program these terms will be derived from *Mie theory* [23].

3. Quaternion Algebra

The tracking of the polarization reference plane can be accomplished using two unit vectors u and v. As the photon scatter through a media, the associated reference plane and vectors rotate in the three dimensional space. Vector rotations can be achieved either using rotational matrices of Euler angles [1, 2] or using quaternion algebra. Quaternions have the advantage of not being susceptible to Gimbal Lock [24]. They were invented by William Hamilton in 1852 and have been used since in a multiplicity of applications, from quantum mechanic, to aeronautical engineering [24], and video game construction [26].

Quaternions are composed of a vector Γ and a scalar q_0. The scalar elements of any quaternion are real, ($q_i \in R^4$), so that a quaternion q is defined as

$$q = q_0 + q_1 i + q_2 j + q_3 k = q_0 + \Gamma \tag{5}$$

where i, j, and k are unit vectors obeying the following properties

$$i^2 = j^2 = k^2 = -1$$
$$ij = k \qquad ji = -k$$
$$jk = i \qquad kj = -i \tag{6}$$
$$ki = j \qquad ik = -j$$

A quaternion can also be expressed as a function of an angle θ

$$q = \cos(\theta) + u\sin(\theta) \tag{7}$$

where $\mathbf{u} = \Gamma/\sin(\theta)$.

The complex conjugate of a quaternion q is

$$q^* = q_0 - q_1 i - q_2 j - q_3 k = q_0 - \Gamma \tag{8}$$

3.1. QUATERNION PROPERTIES

Quaternions multiplication is not commutative as shown in Eq. 7, nevertheless it is very useful because it can be used in rigid body motion algorithms.

3.1.1. *Quaternion Addition and Subtraction*

Two quaternions can be easily added and subtracted by handling their scalars and vector parts separately. Given two quaternions q and t

$$q = q_0 + q_1 i + q_2 j + q_3 k = q_0 + \Gamma$$
$$t = t_0 + t_1 i + t_2 j + t_3 k = t_0 + T \tag{9}$$

their addition is simply

$$q + t = (q_0 + t_0) + (q_1 + t_1)i + (q_2 + t_2)j + (q_3 + t_3)k \tag{10}$$

their subtraction is

$$q - t = (q_0 - t_0) + (q_1 - t_1)i + (q_2 - t_2)j + (q_3 - t_3)k \tag{11}$$

3.1.2. *Quaternion Product*

The dot product of two quaternions is a scalar; the vector product of two quaternions returns a quaternion. The dot product is defined as

$$q \bullet t = t \bullet q = q_0 t_0 + q_1 t_1 + q_2 t_2 + q_3 t_3 \tag{12}$$

While the vector product of two quaternions is

$$h = q \times t = q\,t = h_0 + h_1 i + h_2 j + h_3 k \tag{13}$$

where

$$h_0 = q_0 t_0 - q_1 t_1 - q_2 t_2 - q_3 t_3$$
$$h_1 = q_0 t_1 + q_1 t_0 - q_2 t_3 + q_3 t_2$$
$$h_2 = q_0 t_2 + q_1 t_3 + q_2 t_0 - q_3 t_1$$
$$h_3 = q_0 t_3 - q_1 t_2 + q_2 t_1 + q_3 t_0 \tag{14}$$

The rotation of a vector T about a vector Γ (part of a quaternion q) of an angle 2θ can be achieved multiplying the quaternion formed with the vector Γ and the angle θ by the vector T [25]. The multiplication of the quaternion q to the vector T is possible because the latter can be represented as a quaternion whose scalar term is equal to zero.

$$G = q^{*}Tq \qquad (15)$$

The vector part of the quaternion G resulting from this multiplication is the new rotated vector T.

4. Quaternion Based Monte Carlo Program

4.1. INITIALIZATION AND LAUNCH

The initialization of every Monte Carlo program includes the definition of the photons starting position (for the geometry of Fig. 1, $x = 0$, $y = 0$, and $z = 0$) and beam geometry (isotropic point source, pencil beam, and Gaussian beam are the most common), the initial weight W, and the optical properties of the medium μ_a and μ_s', respectively the absorption and reduced scattering coefficient. For a polarized light Monte Carlo the reference plane must also be defined. Two unit vectors are used for this purpose $v = [v_x, v_y, v_z] = [0, 1, 0]$ and $u = [u_x, u_y, u_z] = [0, 0, 1]$; the vector u replaces the direction cosines commonly used in unpolarized Monte Carlo programs.

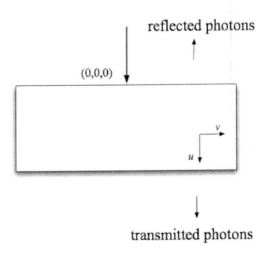

Figure 1. Monte Carlo program geometry at launch.

Finally the initial Stokes vector S is defined; for incident linearly polarized light parallel to the u-v plane $S = [1\ 1\ 0\ 0]^T$.

4.2. HOP – PHOTON MOVEMENT

The step size of the photon in the absorbing and scattering media is expressed using a pseudo random number η and $\mu_t = \mu_a + \mu_s$.

$$s = -\ln\left(\frac{\eta}{\mu_t}\right) \tag{16}$$

The random number is generated in the interval [0 1]. After every step [21], the photon position is updated to a new position

$$[x',\ y',\ z'].$$
$$x' = x + u_x s$$
$$y' = x + u_y s \tag{17}$$
$$z' = x + u_z s$$

4.3. DROP – PHOTON ABSORPTION

The drop step is related to the absorption of light by the media. Every photon launched has a weight W. The weight is reduced after every step of a quantity proportional to the media *albedo* $(a = \mu_s/\mu_s + \mu_a)$, so that after n steps the weight of the photon is $(albedo)^n$. The photon is terminated after reaching a threshold level, if the photon exits the media before reaching such level its corresponding Stokes vector is multiplied by the residual W, to account for photon attenuation.

4.4. SPIN – PHOTON SCATTERING

The scattering section of a Monte Carlo program includes the selection of two angles (the scattering angle α, and the azimuth angle β) that define a new photon trajectory. These angles are used in the rotation of the vectors defining the reference plane, and to update the Stokes vector in the new reference plane.

In polarized Monte Carlo the selection of the angles α and β is accomplished through a process called the *rejection method* [18, 35]. This method is used to generate random variables with a particular distribution using the scatterer phase function.

The phase function for an incident Stokes vector $S=[S_0 \; S_1 \; S_2 \; S_3]^T$) on a particle with the scattering matrix of Eq. 3 is

$$P(\alpha,\beta) = s_{11}(\alpha)S_0 + s_{12}(\alpha)[S_1 \cos(2\beta) + S_2 \sin(2\beta)] \qquad (18)$$

The phase function $P(\alpha, \beta)$ has a bivariate dependence on the angles α and β and is the basis for the construction of the rejection method.

The following steps are necessary for its implementation: first a random angle α is chosen between 0 and π, then a random angle β is chosen between 0 and 2π, and finally a random number η_{rand} between 0 and 1 is generated. $P(0, \beta)$ is used as a normalizing factor to the phase function. This is appropriate since $\alpha = 0$ will always be the angle with the largest scattering intensity.

If $\eta_{rand} \leq P(\alpha, \beta) / P(0, \beta)$ the angle α and β are accepted, and the program can proceed. If $\eta_{rand} > P(\alpha, \beta) / P(0, \beta)$, new α, β, and η_{rand} are randomly selected.

Once the azimuth and scattering angle have been selected the reference plane can be updated through two rotations. This is done through the aforementioned quaternions and the unit vector u and v. The u-v plane rotation is accomplished in two steps illustrated in the figures below.

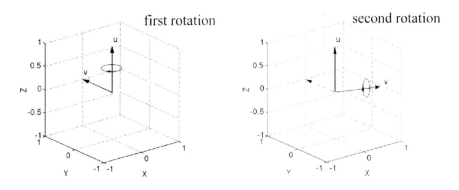

Figure 2. Rotations of the reference plane; first the vector v is rotate about the vector u by an angle β, then the vector u is rotated about the vector v by an angle α.

The vector v is rotated about the vector u by an angle β. As we have shown in section 4.12, the rotations of vectors about other vectors can be achieved through quaternion algebra. A quaternion q_β is first generated.

$$q_\beta = \beta + iu_x + ju_y + ku_z \qquad (19)$$

The operator of Eq. 15 is applied to the vector v. The resulting quaternion is divided into its scalar and vectorial components. The resulting vector is used to repeat the rotation process for the remaining vector u. A new quaternion q_α is created with the scattering angle α and the vector v. The quaternion multiplication $q_\alpha \overset{*}{u} \; q_\alpha$ produces a new quaternion and consequently a new rotated vector u.

The last step of this section is the adjustment of the Stokes vector to the new scattering plane before a scattering event can occur. Every Stokes vector is defined with respect to a reference plane; at the same time a scattering matrix (Eq. 3) is related to a scattering plane. When scattering occurs the photon changes trajectory and the reference plane changes. Hence, before a new scattering event can occur the Stokes vector needs to be *rotated* into the new scattering plane. The rotation of a Stokes vector is achieved with its multiplication to a rotational matrix $R(\beta)$. The Stokes vector can then be multiplied by the appropriate scattering matrix $M(\alpha)$ to obtain the new scattered Stokes vector.

$$S_{scatt} = M(\alpha)R(\beta)S \tag{20}$$

where

$$R(\beta) = \begin{bmatrix} 1 & 0 & 0 & 0 \\ 0 & \cos(2\beta) & \sin(2\beta) & 0 \\ 0 & -\sin(2\beta) & \cos(2\beta) & 0 \\ 0 & 0 & 0 & 1 \end{bmatrix} \tag{21}$$

This process is repeated for every scattering event.

4.5. BOUNDARIES

A photon reaching a boundary is either reflected or transmitted, Fig. 1. The transmitted beam direction will change according to Snell's law. The boundary impact on the Stokes vector is regulated by the Fresnel Mueller matrix, $F_{trans}(\theta)$. Where θ_i is the incident angle θ_t is the transmitted angle and n is the relative index of refraction $n = n_{before\_boundary} / n_{after\_boundary}$.

$$F(\theta_i) = \begin{vmatrix} t_p^2 + t_s^2 & t_p^2 - t_s^2 & 0 & 0 \\ t_p^2 - t_s^2 & t_p^2 + t_s^2 & 0 & 0 \\ 0 & 0 & 2t_p t_s & 0 \\ 0 & 0 & 0 & 2t_p t_s \end{vmatrix} \tag{22}$$

where

$$t_p = 2\cos\theta_i (\cos\theta_t + n\cos\theta_i)^{-1}$$
$$t_s = 2\cos\theta_i (\cos\theta_i + n\cos\theta_t)^{-1}$$

The matrix of Eq. 22 can be used only after the reference plane and the related Stokes vector are oriented so that v is in the refractive plane. This can be done with a simple rotation of an angle ϕ and using the rotational matrix of Eq. 21. The reflection of the photon from a boundary is similarly handled, the coefficient of Eq. 22 in the reflection case are $r_p = nt_p-1$ and $r_s = t_s-1$.

$$S_{trans} = F(\theta_i,\theta_t)R(\phi)S \qquad (23)$$

Stokes vectors enjoy the additive property hence all transmitted and reflected photons can be summed using a polarization sensitive detector. The reference plane has to coincide with the detector reference plane, hence a last $R(\phi)$ rotation of the Stokes vector is necessary.

5. Program Validation

A working version of the program illustrated in the previous sections can be found at ref [26]. The program was tested for a variety of geometries [3,18] and compared to Adding Doubling results [27]. The discrepancy between Monte Carlo and Adding Doubling results was less than 1%. The back-scattered Mueller Matrix for scattering media, such as micro-spheres solutions, is often used as a *gold standard* for polarization sensitive Monte Carlo programs.

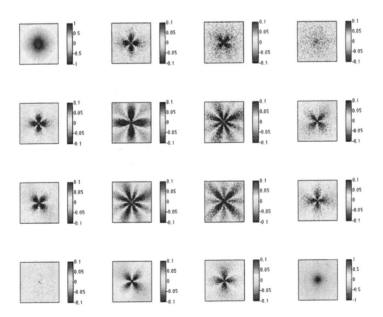

Figure 3. Backscattered Mueller matrix for a solution of 2 μm latex micro-spheres. The index of refraction of the sphere was 1.59. The surrounding media was water n=1.33. Incident wavelength was 0.543 μm.

The images shown in Fig. 3 are the backscattered Mueller matrix images for a solution of 2 µm micro-spheres. The incident beam wavelength was 0.543 µm and the relative index of refraction was 0.75. The first image on the top corresponds to the m_{11} element of the Mueller matrix; the symmetry of all the elements is evident in the images. The simulations shown in Fig. 3 matched the experimental results by Rakovic et al. [16].

6. Conclusions

Polarized sensitive Monte Carlo programs are becoming of common use in the biomedical optics community. Different constructions are possible, and some interesting work has been done in trying to reduce their computational time [28].

Here we have presented an algorithm based on quaternions. The choice of quaternions was dictated by their ease of use, and their immunity to Gimbal Lock, moreover using these elements computational efficiency could be largely improved [24].

A new effort is now under way to make these algorithms more suitable to model biological media. This means including non-spherical scatterers, as well as chiral and birefringent media. More accurate Monte Carlo programs will be indispensable as the use of polarization in the experimental and clinical environment becomes more pervasive.

References

1.   D. Côté and I. Vitkin, "Robust concentration determination of optically active molecules in turbid media with validated three-dimensional polarization sensitive Monte Carlo calculations," Opt. Express 13, 148–163, (2005).
2.   S. Bartel and A. H. Hielscher, "Monte Carlo simulations of the diffuse backscattering Mueller matrix for highly scattering media," Appl. Opt. 39, 1580–1588, (2000).
3.   J. Ramella-Roman, S. Prahl, and S. Jacques, "Three Monte Carlo programs of polarized light transport into scattering media: part I," Opt. Express 13, 4420–4438, (2005).
4.   A. N. Yaroslavsky, V. Neel and R. Rox Anderson "Demarcation of Nonmelanoma Skin Cancer Margins in Thick Excisions Using Multispectral Polarized Light Imaging," Journal of Investigative Dermatology 121, 259–266, (2003).
5.   J. C. Ramella-Roman, K. Lee, S. A. Prahl, and S. L. Jacques, "Design, testing and clinical studies of a hand-held polarized light camera," Journal of Biomedical Optics, 9, 1305–1310, (2004).

6. V. Backman, M. B. Wallace, L. T. Perelman et al. "Detection of prei nvasive cancer cells," Nature 406: 35–36, (2000).

7. C. E. Saxer, J. F. de Boer, B. H. Park, Y. Zhao, Z. Chen, and J. S. Nelson, "High-speed fiber based polarization-sensitive optical coherence tomography of *in vivo* human skin," Opt. Lett. 25, 1355–1357, (2000).

8. G. W. Kattawar and G. N. Plass "Radiance and polarization of multiple scattered light from haze and clouds," Appl. Opt. 7, 1519–1527, (1967).

9. G. W. Kattawar and G. N. Plass, "Degree and direction of polarization of multiple scattered light. 1: Homogeneous cloud layers," Appl. Opt. 11, 2851–2865, (1972).

10. P. Bruscaglione, G. Zaccanti, and W. Qingnong "Transmission of a pulsed polarized light beam through thick turbid media: numerical results," Appl. Opt. 32, 6142–6150, (1993).

11. S. Bianchi, A. Ferrara, and C. Giovannardi, "Monte Carlo simulations of dusty spiral galaxies: extinction and polarization properties," American Astronomical Society 465, 137–144, (1996).

12. S. Bianchi, Estinzione e polarizzazione della radiazione nelle galassie a spirale, Tesi di Laura (in Italian), 1994.

13. S. Martinez and R. Maynard, "Polarization Statistics in Multiple Scattering of light: a Monte Carlo approach," in *Localization and Propagation of classical waves in random and periodic structures* (Plenum Publishing Corporation New York, 1993).

14. S. Martinez, Statistique de polarization et effet Faraday en diffusion multiple de la lumiere Ph.D. Thesis (in French and English), 1984.

15. J. M. Schmitt, A. H. Gandjbakhche, R. F. Bonner, "Use of polarized light to discriminate short-path photons in a multiply scattering medium," Appl Optics; 32, 6535–6546, (1992).

16. M. J. Rakovic, G. W. Kattawar, M. Mehrubeoglu, B. D. Cameron, L. -H. Wang, S. Rastegar, and G. L. Cote, "Light backscattering polarization patterns from turbid media: theory and experiment," Appl. Opt. 38, 3399–3408, (1999).

17. X. Wang and L. V. Wang, "Propagation of polarized light in birefringent turbid media: A Monte Carlo study," J. Biomed. Opt. 7, 279–290, (2002).

18. J. C. Ramella-Roman, S. A. Prahl, and S. L. Jacques, "Three Monte Carlo programs of polarized light transport into scattering media: part II," Opt. Express 13, 10392–10405, (2005).

19. M. Xu, "Electric field Monte Carlo simulation of polarized light propagation in turbid media", Opt. Express 26, 6530–6539, (2004).

20. N. Metropolis and S. Ulam, "The Monte Carlo method," J. Am. Stat. Assoc. 44, 335–341, (1949).

21. L. H. Wang, S. L. Jacques, and L. -Q. Zheng, "MCML – Monte Carlo modeling of photon transport in multi-layered tissues," Computer Methods and Programs in Biomedicine 47, 131–146, (1995).

22. E. Hecht, Optics 4th ed, Pearson Addison Weasley Ed. 2002.

23. Bohren and D. R. Huffman, *Absorption and scattering of light by small particles*, (Wiley Science Paperback Series, 1998).

24. K. Shoemake, "Animating rotation with quaternion curves," Computer Graphics 19, 245–254, (1985).

25. J. J. Craig, *Introduction to robotics. Mechanics and controls,* (Addison-Weseley Publishing Company, 1986).
26. http://faculty.cua.edu/ramella/page2/MonteCarlo/index.html.
27. K. F. Evans and G. L. Stephens, "A new polarized atmospheric radiative transfer model," J. Quant. Spectrosc. Radiat Transfer. 46, 413–423, (1991).
28. F. Jaillon and H. Saint-Jalmes "Description and time reduction of a Mnte Carlo code to simulate propagation of polarized light trhough scattering media." Appl Optics 42, 16 (2003).

A FINITE-DIFFERENCE TIME-DOMAIN MODEL OF OPTICAL PHASE CONTRAST MICROSCOPE IMAGING

STOYAN TANEV*
*Department of Systems and Computer Engineering,
Carleton University, 1125 Colonel By Drive, Ottawa, ON,
Canada K1S 5B6*

JAMES POND AND PAUL PADDON
*Lumerical Solutions, Inc., 660 - 789 West Pender Avenue,
Vancouver BC, Canada V6C 1H2*

VALERY V. TUCHIN,
*Institute of Optics and Biophotonics, Saratov State University,
Saratov, 410012 Russia*

Abstract. The Finite-Difference Time-Domain (FDTD) modeling technique is applied to build a simple simulation equivalent of an optical phase contrast microscope. The model is validated by demonstrating the effect of optical immersion on the optical phase contrast microscope image of a simple biological cell containing a cytoplasm, a nucleus and a membrane. To the best of our knowledge, this is the first study using the FDTD approach to construct optical phase contrast microscope images. The results demonstrate the potential of the FDTD modelling approach and extend its area of applicability a new biomedical research area.

Keywords: Biological cell, finite-difference time-domain method, optical phase contrast microscope, optical clearing effect

1. Introduction

Biomedical photonics has attracted the attention of many researchers from various multidisciplinary fields.[1,2] In many cases, the application of optical software simulation and modeling tools provides a deeper understanding of

---

* To whom correspondence should be addressed. e-mail:tanev@sce.carleton.ca

W.J. Bock et al. (eds.), Optical Waveguide Sensing and Imaging, 243–257.

newly developed optical diagnostics and imaging techniques. The numerical modeling of light interaction with and scattering from biological cells and tissues within the context of a particular optical imaging technique has been of particular interest. There are a number of numerical modeling approaches that can be used for the modeling of the light scattering from biological cells[3] including the finite-difference time domain (FDTD) method.[4–9] The FDTD approach is a powerful tool in studying the nature of the light scattering mechanisms from and imaging of both normal and pathological cells and tissues.

Recently we demonstrated the applicability of the FDTD approach to numerically study how the optical immersion technique (OIT) enhances the effect of cell membrane thickness and gold nanoparticles on optical phase contrast microscopic imaging of cancerous and non-cancerous cells.[7,8] The OIT is based on the so called "optical clearing" (OC) effect consisting in the increased light transmission through microbiological objects due to the matching of the refractive indices (RI) of some of their morphological components to that of the extra-cellular medium.[10–12] In particular, we have studied the changes in the forward scattered light phase and intensity distributions due to different cell membrane thicknesses and different configurations of gold nanoparticles with and without the OIT. Here we apply the FDTD modeling technique to build a simple simulation model of an optical phase contrast microscope. The model will provide the basis for future research studies and additional insights on the OIT benefits for the enhancement of optical phase contrast microscope imaging.

2.   The FDTD Approach

The FDTD technique is a numerical solution of Maxwell's equations.[14] In a source free dielectric medium Maxwell's equations have the form:

$$\nabla \times \mathbf{E} = -\mu_0 \frac{\partial \mathbf{H}}{\partial t}, \tag{1a}$$

$$\nabla \times \mathbf{H} = \varepsilon_0 \varepsilon \frac{\partial \mathbf{E}}{\partial t}, \tag{1b}$$

where E and H are the electric and magnetic fields, respectively, μ_0 is the vacuum permeability and $\varepsilon_0 \varepsilon$ is the permittivity of the medium. Assuming a harmonic ($\propto \exp(-i\omega t)$) time dependence of the electric and magnetic fields and a complex value of the relative permittivity $\varepsilon = \varepsilon_r + i\varepsilon_i$ transforms equation (1b) in the following way:

$$\nabla \times \mathbf{H} = \varepsilon_0 \varepsilon \frac{\partial \mathbf{E}}{\partial t} \Leftrightarrow \nabla \times \mathbf{H} = \omega \varepsilon_0 \varepsilon_i \mathbf{E} + \varepsilon_0 \varepsilon_r \frac{\partial \mathbf{E}}{\partial t} \Leftrightarrow$$
$$\frac{\partial (\exp(\tau) \mathbf{E})}{\partial t} = \frac{\exp(\tau)}{\varepsilon_0 \varepsilon_r} \nabla \times \mathbf{H} \tag{2}$$

where $\tau = \omega \varepsilon_i / \varepsilon_r$ and ω is the angular frequency of the light wave. The continuous coordinates (x, y, z, t) are replaced by discrete spatial and temporal points: $x_i = i \Delta s$, $y_j = j \Delta s$, $z_k = k \Delta s$, $t_n = n \Delta t$, where $i = 0, 1, 2, \ldots, I$; $j = 0, 1, 2, \ldots, J$; $k = 0, 1, 2, \ldots, K$; $n = 0, 1, 2, \ldots, N$. Δs and Δt denote the cubic cell size and time increment, respectively. Using central difference approximations for the temporal derivatives over the time interval $[n\Delta t, (n+1)\Delta t]$ gives

$$\vec{E}^{n+1} = \exp(-\tau \Delta t) \vec{E}^n + \exp(-\tau \Delta t / 2) \frac{\Delta t}{\varepsilon_0 \varepsilon_r} \nabla \times \vec{H}^{n+1/2}, \tag{3}$$

where the electric and the magnetic fields are calculated at alternating half-time steps. The discretization of Eq. (1a) over the time interval $[(n-1/2)\Delta t, (n+1/2)\Delta t]$ (one half time step earlier than the electric field) ensures second-order accuracy of the numerical scheme. In three dimensions, equations (1–3) involve all six electromagnetic field components: E_x, E_y, E_z and H_x, H_y, and H_z. For example, the discretized equations for the H_x and the E_x components take the form:

$$H_x^{n+1/2}(i, j+1/2, k+1/2) = H_x^{n-1/2}(i, j+1/2, k+1/2) + \frac{\Delta t}{\mu_0 \Delta s} \tag{4a}$$
$$\times [E_y^n(i, j+1/2, k+1) - E_y^n(i, j+1/2, k) + E_z^n(i, j, k+1/2) - E_z^n(i, j+1, k+1/2)],$$

$$E_x^{n+1}(i+1/2, j, k) = \exp[-\frac{\varepsilon_i(i+1/2, j, k)}{\varepsilon_r(i+1/2, j, k)} \omega \Delta t] E_x^n(i+1/2, j, k) +$$
$$\exp[-\frac{\varepsilon_i(i+1/2, j, k)}{\varepsilon_r(i+1/2, j, k)} \omega \Delta t / 2] \frac{\Delta t}{\varepsilon_0 \varepsilon_r(i+1/2, j, k) \Delta s} \tag{4b}$$
$$\times [H_y^{n+1/2}(i+1/2, j, k-1/2) - H_y^{n+1/2}(i+1/2, j, k+1/2)$$
$$+ H_z^{n+1/2}(i+1/2, j+1/2, k) - H_z^{n+1/2}(i+1/2, j-1/2, k)],$$

If the symmetry of the structure under consideration allows, one could drop out one of the spatial dimensions and employ a 2D FDTD model which is much easier to conceptualize. For the sake of simplicity, the 2D FDTD model will be introduced here and used to describe the design methodology. The actual simulations, however, are done in 3D. The specifics of the 3D model will be described in the next section.

If the propagation is considered within the framework of a 2D model (x-y plane), Maxwell's equations split into two independent sets as follows: *transverse electric* (TE), with non-zero field components E_x, E_y, H_z, and *transverse magnetic* (TM), with field components H_x, H_y, E_z. In a Cartesian grid system the TE equations for the *x* components of the electric and magnetic fields take the form

$$H_z^{n+1/2}(i,j) = H_z^{n-1/2}(i,j) - \frac{\Delta t}{\mu_0 \Delta s}$$

$$\times [E_y^n(i+1/2,j) - E_y^n(i-1/2,j) + E_x^n(i,j-1/2,k) - E_x^n(i,j+1/2)], \quad (5a)$$

$$E_x^{n+1}(i,j+1/2) = \exp[-\frac{\varepsilon_i(i,j+1/2)}{\varepsilon_r(i,j+1/2)}\omega\Delta t]E_x^n(i,j+1/2) +$$

$$\exp[-\frac{\varepsilon_i(i,j+1/2)}{\varepsilon_r(i,j+1/2)}\omega\Delta t/2]\frac{\Delta t}{\varepsilon_0\varepsilon_r(i,j+1/2)\Delta s} \quad (5b)$$

$$\times [H_z^{n+1/2}(i,j+1) - H_z^{n+1/2}(i,j)],$$

$$E_y^{n+1}(i+1/2,j) = \exp[-\frac{\varepsilon_i(i+1/2,j)}{\varepsilon_r(i+1/2,j)}\omega\Delta t]E_y^n(i+1/2,j) -$$

$$\exp[-\frac{\varepsilon_i(i+1/2,j)}{\varepsilon_r(i+1/2,j)}\omega\Delta t/2]\frac{\Delta t}{\varepsilon_0\varepsilon_r(i+1/2,j)\Delta s} \quad (5c)$$

$$\times [H_z^{n+1/2}(i+1,j) - H_z^{n+1/2}(i,j)],$$

The numerical stability of the FDTD scheme [14] is ensured through the Courant-Friedrichs-Levy condition: $c\Delta t \le (1/\Delta x^2 + 1/\Delta y^2)^{-1/2}$, where c is the light speed in the host medium and Δx and Δy are the spatial steps in the x and y, respectively. The schematic positions of the magnetic and electric field components in a TE FDTD cubic cell are shown in Fig. 1 (left).

In light scattering simulation experiments one uses the so-called total-field/scattered-field (TFSF) formulation[8,14] to excite the magnetic and electric fields and simulate a linearly polarized plane wave propagating in a finite region of a homogeneous absorptive dielectric medium. A schematic representation of the 2D FDTD computational domain is shown in Fig. 1 (right). Total-field scattered-field sources are used to separate the computation region into two distinct regions. The first (dashed line rectangle in Fig. 1 (right) contains the total <u>near</u> fields $E_{tot} = E_{inc} + E_{scat}$ and $H_{tot} = H_{inc} + H_{scat}$, i.e. the sum of the incident field and the scattered near field. The second region (outside of the dashed line rectangle) contains only the scattered <u>near</u> fields $E_{scat} = E_{tot} - E_{inc}$ and $H_{scat} = H_{tot} - H_{inc}$.

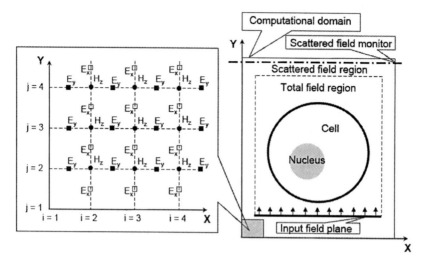

Figure 1. Positions of the electric- and the magnetic-field components in an elementary cubic cell of the 2D TE FDTD lattice (left). Schematic representation of the FDTD computational domain (right).

The TFSF source is particularly useful to study the scattering behavior of objects, as the scattered near field data can be isolated from the incident field and further post-processed to get as much useful information as possible. In Figure 1 (right) a TFSF ideal plane wave is injected along the lower edge of the total field region in the upward direction. Everything inside the TFSF boundary (the dashed line rectangle) is total near field, while everything outside is only scattered near field. Thus in the absence of any objects, the wave propagates within the TFSF region and is subtracted out at the other end, this of course resulting in no scattered field at all. However, if one places an object in the path of the TFSF plane wave, this introduces scattered fields that then propagate outside the total field area. Consequently, one can then measure the total or scattered transmission, by placing monitors inside or the outside of total field region, respectively. Since these are the near fields that are measured, an additional analysis of the results usually requires a post-processing numerical procedure to transform the near fields into their far field counterparts. At the edges, the entire FDTD computational domain is truncated by the so called perfectly matched layer (PML) boundary conditions.[14] The FDTD simulations provide the fields in both, the total and scattered field, regions including the transverse distribution of the forward scattered light from the scattering object which is of high relevance for the modeling optical phase contrast imaging.

3. FDTD 3D Simulation Geometry

The 3D case of the TFSF formulation described above is more complex to implement but follows a very similar logic.[14] The 3D simulation results provided here are based on a modified 3D TFSF formulation that could be more appropriately called "TFRF" (total field/reflected field). The 3D "TFRF" formulation uses a TFSF region which is extended beyond the limits of the simulation domain (Figure 2, left). The extension of transverse dimension of the input field beyond the limits of the computational domain through the PML boundaries would lead to distortions of its ideal plane wave shape[14] which may influence the final simulation results. To avoid these distortions of the input field, periodic boundary conditions must be used. Periodic boundary conditions allow for the simulation of large arrays of periodic features by only simulating a single unit period of the larger array. In our case what that really means is that the actual modeled structure is a periodical row of biological cells. This, however, is not a problem since we are interested in the near scattered field at locations where the coupling effect due to waves scattered from adjacent cells in the row is negligible. This effect can be effectively minimized or completely removed by controlling the lateral dimension of computational domain by using a large enough period of the periodical cell structure. The larger this period, the lower is the coupling effect.

If the plane wave is incident at an angle, which is exactly our case, then Bloch boundaries must be used (Figure 2, right). Bloch boundary conditions

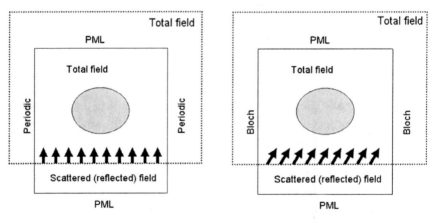

Figure 2. Visual representation of the 3D FDTD simulation approach using periodic (left) and Bloch boundary conditions (right).

are used predominantly in satiations where the simulation requires launching of a plane wave at an angle to a periodic structure. Bloch boundary conditions are periodic boundary conditions which take into account the phase effects due to the tilting of the input plane wave.

In the 3D TFRF formulation the location in the computational domain corresponding to the forward scattered light is positioned within the total field region. The FDTD modeling of optical phase contrast microscopy is based on the phase distribution of the forward scattered field. The requirement for an explicit availability of the forward scattered fields leads to the necessity to perform two different simulations – one with the scattering object and a second one without the object – and providing the distribution of the input field and the total near field in a transverse plane behind the object. The subtraction of the input field from the total field will then provide the scattered field. The phase of the scattered field accumulated by a plane wave propagating through a biological cell will be used in the FDTD model of the optical phase contrast microscope described in the next section.

4.   Optical Phase Contrast Microscopy

Phase contrast microscopy is a contrast-enhancing optical technique that can be utilized to produce high-contrast images of transparent specimens such as microorganisms, thin tissue slices, living cells and sub-cellular components such as nuclei and other organelles. Living cells are characterized by the small difference in the transparency between the structure being imaged and the surrounding medium. In these cases, conventional bright field microscopy often fails to create an image with sufficient contrast. The optical phase contrast technique translates small variations in the phase into corresponding changes in amplitude visualized as differences in image contrast. One of the major advantages of phase contrast microscopy is that living cells can be examined in their natural state without previously being killed, fixed, and stained. As a result, the dynamics of ongoing biological processes can be observed and visualized.[a]

A standard phase contrast microscope design is shown in Figure 3, where an image with a strong contrast ratio is created by coherently re-interfering a reference, or surround beam (S), with a diffracted beam (D) from the specimen. The index contrast between the specimen and the surrounding

---

[a] http://www.microscopyu.com/articles/phasecontrast/phasemicroscopy.html

medium creates a phase delay between the two beams, which can be augmented by a phase plate in the S beam. The resulting image, where the phase difference is translated into an amplitude variation by interference at the image plane, can have a high contrast ratio, particularly if both beams have the same amplitude at the image plane. Figure 3 also illustrates the part of the microscope that will become the subject of FDTD modeling combined with Fourier optics.

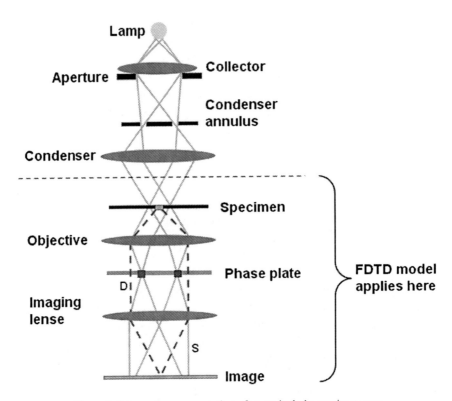

Figure 3. Schematic representation of an optical phase microscope.

Figure 4 provides a 2D visual representation illustrating the major steps in the FDTD model. It could be associated with the 2D FDTD model described above. The phase contrast microscope uses incoherent annular illumination that could be approximately modeled by four different simu- lations with two plane waves incident at a given polar angle and two different polarizations (one in and one perpendicular to the propagation plane and corresponding to the TE and TM cases described above).

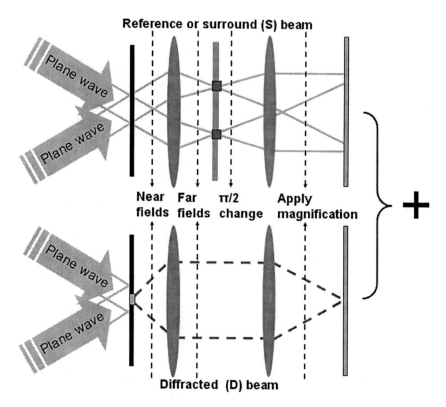

Figure 4. 2D visual representation of the optical phase contrast microscope FDTD model using incoherent illumination by two planes waves at polar angles of 30 degrees. For each of the two plane waves light propagation is modeled as a combination of two parallel wave phenomena: a) propagation of the reference (surround) beam without the scattering object, and b) propagation of the diffracted beam which is due to the scattering object. The phase contrast image is created by adding up incoherently (in intensity) the two final interference images corresponding to the two input planes waves.

5. FDTD 3D Simulations

5.1. CELL STRUCTURE

The simulation results presented here are based on a 3D FDTD model. The cell is modeled as a dielectric sphere with a radius $R_c = 1.5$ μm and membrane thickness $d = 10$ nm. The cell nucleus is also spherical, has a radius $R_n = 0.5$ μm and its center is located at a position which is 0.5 μm shifted from the cell center in a direction perpendicular to the principal direction of propagation. The refractive index of the cytoplasm is $n_c = 1.36$, of the nucleus $n_n = 1.4$, of the membrane $n_m = 1.47$ and of the extra-cellular material n_e = 1.33 (no refractive index matching) or 1.36 (refractive index matching).

The FDTD simulation uses non-uniform meshing where the number of mesh points in space is automatically calculated to ensure a higher number of mesh points in materials with higher values of the refractive index.[15] The time step used during the simulation was defined by means of a factor of 0.99 in the Courant stability limit: $c\Delta t = 0.99 \times \left(1/\Delta x^2 + 1/\Delta y^2 + 1/\Delta z^2\right)^{-1/2}$.

5.2. FDTD NEAR FIELDS

The annular illumination of the phase contrast microscope is modeled in an approximate way by running 8 separate simulations each involving an input plane wave incident at a polar angle of 30 degrees, an azimuthal angle, a specific polarization and wavelength λ = 632 nm. The illumination scheme can be visually conceptualized by means of Fig. 3, where the two input plane waves would represent 4 out of the 8 cases corresponding to two opposite azimuthal angles and two polarizations – one in the plane of and one perpendicular to the plane of the figure. The other 4 cases correspond to the two input planes associated with the two azimuthal angles in the plane perpendicular to the plane of the figure. Each of the 8 incident plane waves (defined by its polar angle – 30 degrees, azimuthal angles – 0, 90, 180 or 270 degrees, and polarization – 0 or 90 degrees) is simulated separately to propagate and scatter off the cell. Every single FDTD simulation provides the near field components corresponding to the scattered (D) and the reference (S) beams in a transverse plane located right after the scattering object – the cell (see Fig. 3).

5.3. FAR FIELD TRANSFORMATION

The far field transformations (Figure 5) use the FDTD calculated near fields right after the cell and return the three complex components of the electro-magnetic fields at 1 meter distance (long enough) from the location of the near fields, i.e. in the far field. To scale the far field to a different distance we can recognize that in 3D the electric field intensity decreases like $1/R^2$: $|E(R)|^2 = |E_0|^2/R^2$, where E_0 is the field at a distance of 1 m as returned by the far field transformation and R is measured in meters. The details of the far field transformation are described by Taflove in his classic book on FDTD.[14]

The end result of the far field transformation is: $E_r(u_x, u_y)$, $E_\theta(u_x, u_y)$ and $E_\Phi(u_x, u_y)$ where (r, θ, Φ) refer to a spherical coordinate system and the variables u_x and u_y are the x and y components of the unit direction vector **u**. The unit direction vector is related to the angular variables by, $u_x = \sin(\theta)\cos(\Phi)$, $u_y = \sin(\theta)\sin(\Phi)$, $u_z = \cos(\theta)$, with $u_x^2 + u_y^2 + u_z^2 = 1$. The in-plane wave vectors for each plane wave are given by $k_x = k\, u_x$ and $k_y = k\, u_y$, where $k = 2\pi/\lambda$. Now that we have the components of $E(k_x, k_y)$ we can use their amplitudes and phases to determine all the properties of the polarization as well as do Fourier optics for both the scattered and reference beams.

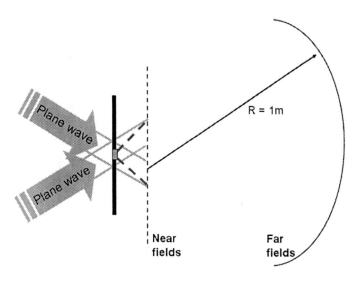

R = 1m

Plane wave

Plane wave

Near
fields

Far
fields

Figure 5. Visual representation of the far field transformation procedure within the context of the optical phase contrast microscope schematics presented in Figures 3 and 4.

5.4. OPTICAL MAGNIFICATION

The magnification factor is applied on the far fields (see Figure 4) before the interference of the scattered (S) and reference (D) beams. A lens system that provides magnification is only modifying the angle of light propagation. So it is as simple as multiplying the direction cosines, u_x and u_y by the inverse value of the desired magnification factor M: $U_x = u_x/M$ and $U_y = u_y/M$. Normally, this would lead to a lot of complications because of the vectorial nature of the E field. However, the beauty of working in spherical coordinates (E_r, E_θ and E_Φ) is that the vectorial components do not change when u_x and u_y are modified because they are part of a local coordinate system that is tied to the values of u_x and u_y. In our particular case the image is reconstructed assuming a $10 \times$ magnification.

5.5. NUMERICAL APERTURE

It is also possible to apply the effect of a numerical aperture which clips any light that has too steep an angle, i.e. would not be collected by the lens system. This means that all beams with $U_x^2 + U_y^2 > NA^2$ are being clipped. In our case $NA = 0.8$. The effect of the aperture will be defined by the inequality $U_x^2 + U_y^2 > NA^2$ applied to the corrected aperture angles θ' and Φ':

$$\sin(\Phi') = U_y/U_{xy}, \cos(\Phi') = U_x/U_{xy}, \cos(\theta') = U_z, \sin(\theta') = U_{xy},$$

where $U_{xy} = sqrt(U_x^2 + U_y^2)$, $U_z = sqrt(1 - U_{xy}^2)$ and the "sqrt" labels a square root mathematical operation. The magnified field components will then have the following form:

Scattered (D) beam:
$$E_x(k_x,k_y) = -E_\Phi \sin(\Phi') + E_\theta \cos(\Phi')\cos(\theta'),$$
$$E_y(k_x,k_y) = E_\Phi \cos(\Phi') + E_\theta \sin(\Phi')\cos(\theta'),$$
$$E_z(k_x,k_y) = -E_\theta \sin(\theta'),$$

Reference (S) beam:
$$E_{x-ref}(k_x,k_y) = -E_{\Phi-ref} \sin(\Phi') + E_{\theta-ref}\cos(\Phi')\cos(\theta'),$$
$$E_{y-ref}(k_x,k_y) = E_{\Phi-ref}\cos(\Phi') + E_{\theta-ref}\sin(\Phi')\cos(\theta'),$$
$$E_{z-ref}(k_x,k_y) = -E_{\theta-ref}\sin(\theta'),$$

where $E_{\theta-ref}$ are $E_{\Phi-ref}$ are the far field components of the incident beam (calculated as if propagated in the absence of the scattering object).

The fields given above are then used to calculate back the Fourier inverse transform of the far field transformed fields leading to the distribution of the scattered and the references beams in the image plane:

Scattered (D) beam:
$$E_{x\_image} = sum(E_x(k_x,k_y) \exp(ik_x x + ik_y y + ik_z z)),$$
$$E_{y\_image} = sum(E_y(k_x,k_y) \exp(ik_x x + ik_y y + ik_z z)),$$
$$E_{z\_image} = sum(E_z(k_x,k_y) \exp(ik_x x + ik_y y + ik_z z)),$$

Reference (S) beam:
$$E_{x\_ref\_image} = sum(E_{x-ref}(k_x,k_y) \exp(ik_x x + ik_y y + ik_z z)),$$
$$E_{y\_ref\_image} = sum(E_{y-ref}(k_x,k_y) \exp(ik_x x + ik_y y + ik_z z)),$$
$$E_{z\_ref\_image} = sum(E_{z-ref}(k_x,k_y) \exp(ik_x x + ik_y y + ik_z z)),$$

where the summation is over all angles. The intensities of the two images at the image plane will be in the form:

$$I_{image} = abs(E_{x\_image})^2 + abs(E_{y\_image})^2 + abs(E_{z\_image})^2$$
$$I_{ref\_image} = abs(E_{x\_ref\_image})^2 + abs(E_{y\_ref\_image})^2 + abs(E_{z\_ref\_image})^2$$

No refractive index matching   **Refractive index matching**

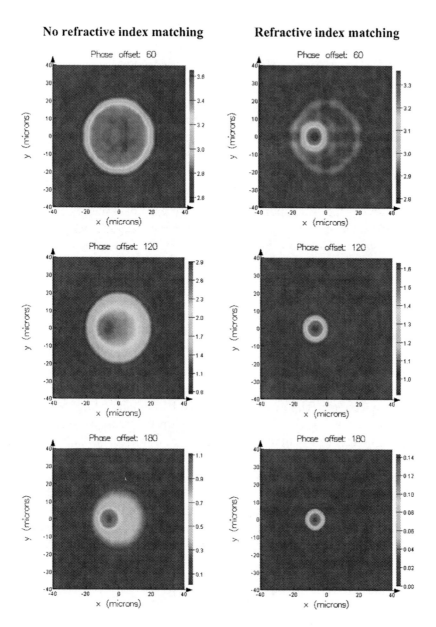

Figure 6. FDTD simulation calculated optical phase contrast images of a cell with a nucleus and membrane. The value of the phase offset corresponds to the phase difference Ψ between the diffracted (D) and reference (S) beams.

5.6. IMAGE CREATION

The calculation described so far is done with the scattered beam and a reference beam, so the two can be added with any desired phase offset Ψ:

$$
\begin{aligned}
I = \ &abs(E_{x\_image} + a\,E_{x\_ref\_image}\,\exp(i\Psi))^2 \\
&+\ abs(E_{y\_image} + a\,E_{y\_ref\_image}\,\exp(i\Psi))^2 \\
&+\ abs(E_{z\_image} + a\,E_{z\_ref\_image}\,\exp(i\Psi))^2.
\end{aligned}
$$

The coefficient a and the phase Ψ correspond to the ability to adjust the relative amplitudes and the phase difference between the two beams.

6. FDTD Results

Figure 6 shows the simulation results as false grayscale images for a given phase difference Ψ between the reference beam and the scattered beam (assuming $a = 1$) for two different cases: refractive index matching (right) and no refractive index matching (left). It can be seen that subtle changes in the refractive index matching greatly improve the optical phase contrast microscope image.

7. Conclusions

We have provided preliminary results on the application of the FDTD approach to the development of a simulation equivalent of an optical phase contrast microscope. The model is validated by demonstrating the effect of optical immersion on the optical phase contrast microscope images of a simple biological cell containing a cytoplasm, a nucleus and a membrane. The validation of the model provides a basis for future studies of optical phase contrast microscope imaging involving effects such as the presence of gold nanoparticles and nonlinear optical phenomena. The possibility to process FDTD simulation results in providing real life biological cell images was found to be very promising. The results demonstrate the potential of the FDTD modelling approach and extend its area of applicability into a new biomedical research area.

Acknowledgments

ST, JP and PP acknowledge the use of the computing resources of WestGrid (Western Canada Research Grid) – a $50-million project to operate grid-enabled high performance computing and collaboration infrastructure at institutions across Canada. VVT was supported by grants of Federal Agency of Education of RF № 1.4.06, RNP.2.1.1.4473 and by CRDF BRHE grant RUXO-006-SR-06.

References

1. P. N. Prasad, "Bioimaging: principles and techniques", Chap. 7 in *Introduction to Biophotonics*, pp. 203–249, John Wiley & Sons, New Jersey (2003).
2. V. V. Tuchin, *Tissue Optics: Light Scattering Methods and Instruments for Medical Diagnosis*, Second edition, PM 166, Bellingham, WA, USA, 2007.
3. F. M. Kahnert, "Numerical methods in electromagnetic scattering theory," *Journal of Quantitative Spectroscopy and Radiative Transfer* 73, 775–824 (2003).
4. R. Drezek, A. Dunn and R. Richards-Kortum, "A pulsed finite-difference time-domain (FDTD) method for calculating light scattering from biological cells over broad wavelength ranges," *Optics Express* 6, 147–157 (2000).
5. T. Tanifuji and M. Hijikata, "Finite difference time domain (FDTD) analysis of optical pulse responses in biological tissues for spectroscopic diffused optical tomography," *IEEE Transactions on Medical Imaging* 21, 181–184 (2002).
6. R. Drezek, M. Guillaud, T. Collier, I. Boiko, A. Malpica, C. Macaulay, M. Follen, R. R. Richards-Kortum, "Light scattering from cervical cells throughout neoplastic progression: influence of nuclear morphology, DNA content, and chromatin texture," *Journal of Biomedical Optics* 8, 7–16 (2003).
7. S. Tanev, V. V. Tuchin and P. Paddon, "Light scattering effects of gold nanoparticles in cells," *Laser Physics Letters*, Vol. 3, No. 12, 594–598 (2006).
8. S. Tanev, V. V. Tuchin and P. Paddon, "Cell membrane and gold nanoparticles effects on optical immersion experiments with noncancerous and cancerous cells: finite-difference time-domain modeling", *Journal of Biomedical Optics*, Vol. 11, No. 6, 064037 (2006).
9. X. Li, A. Taflove, and V. Backman, "Recent progress in exact and reduced-order modeling of light-scattering properties of complex structures," *IEEE J. Selected Topics in Quantum Electronics* 11, 759–765 (2005).
10. R. Barer, K. F. A. Ross, and S. Tkaczyk, Refractometry of living cells, *Nature* 171 (1953), pp. 720–724.
11. B. A. Fikhman, *Microbiological Refractometry*, Medicine, Moscow (1967).
12. V. V. Tuchin, *Optical Clearing of Tissues and Blood*, Vol. PM 154, SPIE Press, 2005.
13. W. Sun, N. G. Loeb, S. Tanev, and G. Videen, "Finite-difference time-domain solution of light scattering by an infinite dielectric column immersed in an absorbing medium," *Applied Optics* 44, 1977–1983 (2005).
14. A. Taflove and S. Hagness, *Computational Electrodynamics: The Finite-Difference Time Domain Method*, Artech House Publishers (3rd edition), Boston (2005).
15. The simulations were performed by the FDTD Solutions™ software developed by Lumerical Solutions Inc., Vancouver, BC, Canada: www.lumerical.com

OUT OF PLANE POLARIMETRIC IMAGING OF SKIN: SURFACE AND SUBSURFACE EFFECT

JESSICA C. RAMELLA-ROMAN*
The Catholic University of America, 620 Michigan Avenue NE, Washington, DC 20064

Abstract. True borders of certain skin cancers are hard to detect by the human eye. For this reason, techniques such as polarized light imaging have been used to enhance skin cancer contrast before Mohs surgery procedures. In standard polarized light imaging the effect of the exposed rough surface is minimized using an index-matched boundary, such as a glass slide with gel. Moreover, surface glare is eliminated using indirect illumination. We studied the effect of surface roughness on the polarized light backscattered from skin demonstrating that rough surface effects can be minimized using out-of-plane polarized illumination in conjunction with polarized viewing, without the need for an index-matched boundary.

Keywords: Polarization, skin cancer imaging, rough surface

1. Introduction

Polarized light imaging has been used in dermatology to enhance the borders of certain skin lesions, such as basal cell carcinoma and squamous cell carcinoma [1,2]. A typical set up for polarized light imaging is shown in Fig. 1. A camera views the skin perpendicular to the surface. The source illuminates the skin obliquely about 20° from the surface normal. The incident light is linearly polarized, with its electric field oriented parallel (*p*) to the plane defined by the locations of the source, the illuminated tissue, and the camera.

---

* To whom correspondence should be addressed. Jessica C. Ramella-Roman, Department of 0020 Biomedical Engineering, The Catholic University of America, 620 Michigan Avenue NE, Washington DC, 20064. E mail: ramella@cua.edu

W.J. Bock et al. (eds.), Optical Waveguide Sensing and Imaging, 259–269.
© 2008 *Springer.*

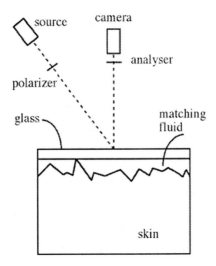

Figure 1. Typical set up for polarized light imaging. The camera is viewing normal to surface.

A glass plate, often with an index-matching gel or fluid, contacts the skin in the image plane to ensure a flat surface. Glare due to specular reflection off the air/glass interface and the glass/skin interface is deflected obliquely and misses the camera. Only light scattered from within the skin is imaged by the camera. Finally, a linear polarization analyzer is placed in front of the camera. Two images are acquired, the first image (I_{pp}) with the analyzer aligned parallel to the scattering plane, and the second image (I_{ps}) with the analyzer aligned perpendicular (s) to the scattering plane.

Photons backscattered from superficial layers largely retain the polarization orientation of the source and contribute more strongly to the I_{pp} image. Light returning from deeper layers tends to be multiply scattered, contributing equally to the I_{ss} and I_{pp} images. A polarized light image P is constructed from

$$P = \frac{I_{pp} - I_{ps}}{I_{pp} + I_{ps}}. \tag{1}$$

By subtracting I_{ps} from I_{pp}, we eliminate the contribution of the multiply scattered photons. By normalizing by $I_{pp} + I_{ps}$, we attempt to cancel variations in the illumination and effects of superficial melanin pigmentation.

The index-matching glass plate guarantees a flat boundary eliminating scattering from the exposed rough skin. Eliminating the glass boundary would be particularly advantageous in a clinical setting since it would eliminate physical contact with the patient. When in-plane illumination is used, as in Fig. 1, without the glass plate, light scattered by the air-skin interface also contributes to the polarized image. In such conditions, it was found that in-plane illumination from different angles could not eliminate this source of scatter [3].

Germer, et al. [3,4] used light scattering ellipsometry to distinguish rough surface from subsurface scatter from a variety of materials, such as silicon, glass, and metals. They found that a number of different single-scattering mechanisms did not depolarize the light, and yielded distinct polarization states that depend upon incident direction and polarization. They further found that contrast between the different mechanisms was greatest when light was incident at an oblique angle with its electric field parallel to the plane of incidence, and when the scattered light was viewed out of that plane. Imaging setups that takes advantage of these findings are able to eliminate the influence of the light scattering from the rough surface, thus highlighting the subsurface scattered light. We show that when using this out-of-plane illumination technique the effect of the rough surface on the images I_{pp} and I_{ps} can be substantially reduced by choosing a source azimuth angle ϕ larger than 50°, a source inclination angle $\theta = 45°$, and a viewing inclination angle $\theta = -45°$.

1.1. THEORY

The scattering of a polarized monochromatic light beam from a surface or a subsurface particle can be expressed in terms of a Jones scattering matrix $\mathbf{q}$,

$$
\begin{bmatrix} E_s^{scat} \\ E_p^{scat} \end{bmatrix} = \frac{\exp(ikR)}{R} \begin{bmatrix} q_{ss} & q_{ps} \\ q_{sp} & q_{pp} \end{bmatrix} \begin{bmatrix} E_s^{inc} \\ E_p^{inc} \end{bmatrix}
\tag{2}
$$

where $k = 2\pi/\lambda$, R is the distance from the scatterer. $\boldsymbol{E}^{scat}$ is the scattered electric field, and $\boldsymbol{E}^{inc}$ is the incident electric field. The development of the theory of rough surface scattering has occurred in the last forty years [5]; we considered two limiting cases: the *smooth surface limit* for rough surfaces, and the Raylegh limit for subsurface scattering. In the smooth surface limit where the root mean square (*rms*) of the surface height function is much smaller than the incident wavelength and the surface slopes are small, Rayleigh-Rice theory can be applied, and the elements of the scattering matrix are proportional to

$$q_{ss}^{sub} = k^2 \cos\phi_r / \left[\left(k_{zi} + k'_{zi}\right)\left(k_{zr} + k'_{zr}\right)\right]$$

$$q_{ps}^{sub} = -kk'_{zi} \sin\phi_r / \left[\left(\varepsilon k_{zi} + k'_{zi}\right)\left(k_{zr} + k'_{zr}\right)\right]$$

$$q_{sp}^{sub} = -kk'_{zr} \sin\phi_r / \left[\left(k_{zi} + k'_{zi}\right)\left(\varepsilon k_{zr} + k'_{zr}\right)\right] \qquad (3)$$

$$q_{pp}^{sub} = \left(\varepsilon k_{xyi} k_{xyr} - k'_{zi} k'_{zr} \cos\phi_r\right) / \left[\left(\varepsilon k_{zi} + k'_{zi}\right)\left(\varepsilon k_{zr} + k'_{zr}\right)\right]$$

where

$$k_{zi} = k\cos\theta_j$$

$$k_{xyj} = k\sin\theta_j \qquad (4)$$

$$k'_{zj} = k\left(\varepsilon - \sin^2\theta_j\right)^{1/2}$$

Germer [6] showed that in the Rayleigh limit for small spheres located below a surface the scattering matrix elements are propor-tional to

$$q_{ss}^{sub} = k^2 \cos\phi_r / \left[\left(k_{zi} + k'_{zi}\right)\left(k_{zr} + k'_{zr}\right)\right]$$

$$q_{ps}^{sub} = -kk'_{zi} \sin\phi_r / \left[\left(\varepsilon k_{zi} + k'_{zi}\right)\left(k_{zr} + k'_{zr}\right)\right]$$

$$q_{sp}^{sub} = -kk'_{zr} \sin\phi_r / \left[\left(k_{zi} + k'_{zi}\right)\left(\varepsilon k_{zr} + k'_{zr}\right)\right] \qquad (5)$$

$$q_{pp}^{sub} = \left(k_{xyi} k_{xyr} - k'_{zi} k'_{zr} \cos\phi_r\right) / \left[\left(\varepsilon k_{zi} + k'_{zi}\right)\left(\varepsilon k_{zr} + k'_{zr}\right)\right]$$

Eqs. 3 and 5 are very similar; only the q_{pp} terms differ and only when θ_i and $\theta_s \neq 0$. That difference is even stronger if $\phi_r \neq 0$. Thus, to distinguish these scattering mechanisms, both the incident and the viewing directions must be oblique, and viewing out of the plane of incidence enhances the distinction. Furthermore, measurement of the polarization state of the scattered light in these geometries for well-chosen incident polarization can be used to unambiguously distinguish them. This technique has been utilized [6] to identify particles and small subsurface defects on semiconductor media, we extend its use to skin roughness measurements.

Certain combinations of incident, viewing directions, and polarizations yield no signal from the roughness. For example, due to q_{ps}^{rough} being zero when $\phi_r = 0$ for any incident angle and p-polarization, there will be no light having s-polarization in the specular direction that, within the limitations of the theory, arises from the rough interface. Another combination occurs when the numerator of q_{pp}^{rough} is zero. For an index of 1.333, for

$\theta_i = \theta_r = 45°$, and for p-polarized incident light, no light will be no p-polarized light scattered with an azimuth angle ϕ_r near 45°.

One metric that has proven to be very sensitive to scattering mechanism is the orientation angle of the polarization ellipse, η, of the scattered light for a given incident polarization state. In terms of Stokes parameters η is given by

$$\eta = \arctan(S_2 / S_1) / 2 \tag{6}$$

where S_1 and S_2 are the second and third elements of the Stokes vector, given by

$$\begin{aligned} S_1 &= I_{ps} - I_{pp} \\ S_2 &= 2I_{p45} - I_{ps} - I_{pp} \end{aligned} \tag{7}$$

When $\eta = 0°$ the light is s-polarized and when $\eta = \pm90°$, it is p-polarized.

The surface roughness of the skin varies depending on location, patient age, and photo-exposure, among many other factors. Several roughness types are also present in close proximity. In the stratum corneum flat corneocytes overlap every 25 μm their thickness is in on the order of 0.5 μm [7]; the corneocytes alone can be considered micro-rough. Glyphic

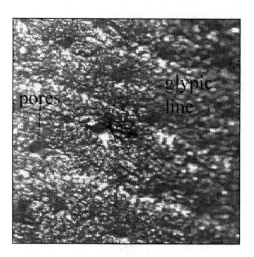

Figure 2. Scattering from the skin. The elements creating specular reflection in this co-polarized image are not the pores or the glyphic lines but something much smaller, possibly the corneocytes.

wrinkles are another source of roughness ranging from less than one millimeter to several millimeters; their size put them in the micro-facet scale [9]. Finally the dermal epidermal junction could contribute to the rough scatter mechanism. The peaks and valley in between the two strata have an *rms* height close to 10 μm and correlation length of circa 80 μm. The roughness of the skin is then a mixture of roughness close to the smooth surface limit and lager roughness better approximated with a facet model. This concept is illustrated in Figure 2.

This image shows a portion of skin images with parallel polarizers to enhance the specular reflection from skin. Skin pores as well as glyphic lines are visible on the skin. Most of the specularly reflected light comes by small elements within the glyphic lines margins possibly corneocytes.

2. Experimental Procedure

Figure 3 shows a schematic of a system used to acquire images as a function of out-of-plane illumination direction. The source is oriented at a fixed angle $\theta = 45°$ with respect to the sample surface normal and is allowed to move about an azimuth angle ϕ from 0° to 140° in steps of 2°.

The source consists of a Xenon arc lamp coupled to a fiber bundle (Oriel, Stratford CT). A short pass filter eliminated infrared radiation while a band pass filter is used to center the wavelength at 543 nm with a 10 nm bandwidth (Newport Corporation, Franklin MA). A dichroic sheet polarizer (Melles Griot, Rochester NY) selects a specific incident linear polarization state, polarizers have extinction coefficient equal to 10^{-4} in the visible range. The polarizer is mounted on a computer-controlled rotational stage (Newport Corporation, Franklin MA). Finally, a lens collimates the light incident

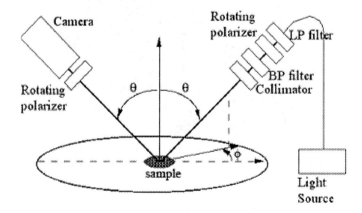

Figure 3. Polarized multi-directional illumination system used in this study.

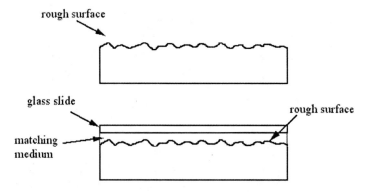

Figure 4. Two sample modalities.

upon the sample (Newport Corporation, Franklin MA). The diameter of the incident beam is about 3 cm.

The optical collection system is a 10-bit 600 × 600 pixel monochrome digital camera (Lumenera, Ottawa Canada) and lens assembly. In front of the camera is another computer-controlled rotating dichroic sheet polarizer to analyze the polarization state of the scattered light. The camera viewes the sample from a fixed orientation with $\theta = 45°$.

Measurements were first performed directly with the rough skin surface exposed, then repeated after the sample was smeared with a moisturizing transparent gel and covered with a round 0.25 mm thick piece of glass.

By matching the air-tissue boundary with the glass, the scattering by the roughness of the air-tissue interface is eliminated.

3. *In vitro* Studies

Figure 5 shows a series of I_{pp} images acquired with and without the index-matching glass plate and under three illumination directions. The three images taken with the glass plate show the edge of the plate and the spreading gel, and so, only the right part of the image should be considered. The three acquired with the glass plate show relatively little structure, since most of the light is diffusely scattered beneath the surface. The three I_{pp} images acquired without the glass plate, in contrast, show significant structure from surface roughness, and the observed structure varies with illumination direction. When $\phi > 45°$ the surface roughness is no longer visible and only single and multiply scattering photon from the subsurface reach the camera. The images with the glass slide and without are very similar.

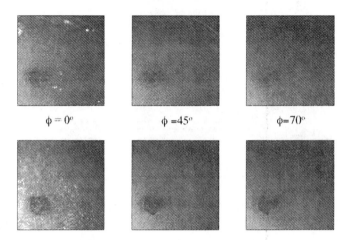

Figure 5. *p*- images acquired with a glass plate (top row) and without a glass plate (bottom row) of a porcine skin sample measured with different illumination azimuths. The size of the imaged sample is approximately 17×17 mm^2. The bottom edge of each image is closer to the camera than the top edge.

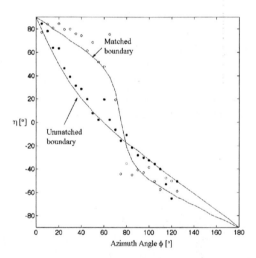

Figure 6. The principal angle of polarization η as a function of the azimuth angle ϕ. The full symbols correspond to a rough surface sample; the line going through the data is a micro-rough surface model. The void symbols are experimental data obtained a sample where the boundary was matched with a glass slide and gel.

The principal angle of polarization, η, can be calculated from a set of polarized. When the incident light is p-polarized, and s-, p-, and 45°-polarized images are captured equation 6 can be used to calculate η. Figure 6 shows the angle η as a function of ϕ measured on samples of porcine skin with and without the matched boundary (gel and glass slide).

There is a significant amount of apparent noise in the data, judging from the variations from point to point, which presumably result from the high degree of depolarization. Nevertheless the data show that when the azimuth angle is above 50 degrees the principal angle of polarization is very similar for both the rough surface measurements and the matched boundary measurements. The rough surface effect is minimized and the glass slide is no longer necessary.

3.1. MODELING

The results related to the rough surface experiment (no glass slide) were modeled with a micro-roughness model in the smooth-surface limit, Figure 6 thin black line, to take into account the scattering of superficial corneocytes. A micro-roughness surface model with a Gaussian power spectral density with *rms height* of 10 μm and correlation length equal to 5 μm is able to follow the data, Figure 6. The index of refraction for skin was set at $n =$ 1.48. The data was modeled using the program Model Integrated Scatter Tool (MIST) [8].

The matched boundary data was also modeled with MIST, using a two source BRDF model. The two source model is composed of a flat single scattering interface with associate reflectance coefficient (5%) and index of refraction $n = 1.48$ and a diffusive media underneath. A Henyey-Greenstein phase function with anisotropy $g = 0.96$, and scattering coefficient $\mu_s = 600$ cm^{-1} was used in this model. The proposed model is able to capture the fast change in η from +40° to –40° at $\phi = 70°$ and although a lot of variability is still present in the data. The standard error for the experimental data is smaller than the used symbols.

4. Conclusions

The light multiply scattered beneath the surface is expected to lose information of its initial polarization state and direction. That is, that fraction of light in an image that results from multiple scattering should be relatively insensitive to incident polarization and direction.

Light scattering from the porcine samples is influenced by the surface roughness as well as the subsurface particles such as cell membranes, nuclei, and collagen bundles; the scattering in the subsurface contributes to maintain a low degree of polarization.

The finding that light scattered by surface roughness has a specific orientation allow one to be insensitive to roughness by viewing only the orthogonal polarization. If one is illuminating the surface with p-polarization and viewing at $\theta_r = 45°$, then when illuminating from $\theta_i = 45°$ and $\phi_i = 0°$, one should view only the p-polarized component, while when illuminating from $\theta_i = 45°$ and $\phi_i = 45°$, one should view only the s-polarized component. The combination of these two images could be used to eliminate the multiply scattered light that should be independent of the incident polarization enhancing single scattering from structures in the surface.

Finally different incident angles can be used to show different roughness effects. Most of the reflected polarized light comes from the rough surface boundary, so the degree of linear polarization Eq. 8 is particularly sensitive to the rough surface response.

$$I_{pol} = \frac{\sqrt{S_1^2 + S_2^2}}{S_o} \tag{8}$$

Figure 7 and 8 show I_{pol} for light incident at the specular angle $\theta_i = 45°$ and $\phi_i = 0°$ and at $\theta_i = 45°$ and $\phi_i = 90°$ as well as the corresponding pp and ps images.

The resulting I_{pol} images are quite different; one enhancing roughness due to hair follicles and skin pores the other more indicative of local surface height differences. The out-of-plane technique is currently being applied to imaging of skin lesions and skin cancer.

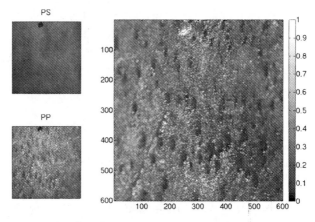

Figure 7. Enhancement of surface features with the degree of linear polarization. The image was collected with angles $\theta_r = 45°$ and $\phi_i = 0°$. Skin pores are very visible both in the I_{pol} image (right) and the pp image (left).

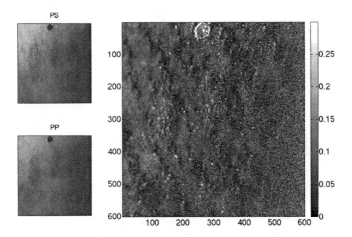

Figure 8. Enhancement of surface features with the degree of linear polarization. The image was collected with angles $\theta_r = 45°$ and $\phi_i = 90°$. In this image skin pores are less visible and different roughness effects stand out. The I_{pol} image (right) and the *pp* image (left).

References

1. R. R. Anderson, "Polarized light examination and photography of the skin." Arch. Dermatol. 127, 1000–1005, (1991)
2. L. Jacques, J. C. Ramella-Roman, K. Lee. "Imaging skin pathology with polarized light," *Journal of Biomedical Optics*, 7, 329–340, (2002).
3. T. A. Germer and C. C. Asmail, "Polarization of light scattered by microrough surfaces and subsurface defects", *J. Opt. Soc. Am. A*, 16, 1326–1332 (1999).
4. S. Saidi, S. L. Jacques, F. K. Tittel, "Mie and Rayleigh modeling of visible-light scattering in neonatal skin," *Applied Optics* 34, 7410–7418, (1995).
5. Beckmann, P. and Spizzichino A. The Scattering of Electromagnetic Waves from Rough Surfaces, Pergamon Press, 1963.
6. T. A. Germer, "Angular dependence and polarization of out-of-plane optical scattering from particulate contamination, subsurface defects, and surface microroughness", *Applied Optics* 36, 8798–8805 (1997).
7. J. Q. Lu, X. H. Hu, K. Dong, "Modeling of the rough-interface effect on a converging light beam propagating in a skin tissue phantom," *Applied Optics* 39, 5890–5896 (2000).
8. T. A. Germer, *SCATMECH: Polarized Light Scattering C++ Class Library* available from http://physics.nist.gov/scatmech.
9. K. E. Torrance E.M. Sparrow, "Theory of off-specular reflections from roughened surfaces," *J. Opt. Soc. Am.*, V. 57, No. 9, 1105–1114, (1967).